FOCUS ON PROBABILITY THEORY

Focus on Probability Theory

Louis R. Velle
Editor

Nova Science Publishers, Inc.
New York

For permission to use material from this book please contact us:
Telephone 631-231-7269; Fax 631-231-8175
Web Site: http://www.novapublishers.com

NOTICE TO THE READER

LIBRARY OF CONGRESS CATALOGING-IN-PUBLICATION DATA
Available upon request

ISBN 1-59454-474-3

Published by Nova Science Publishers, Inc. ✢ New York

CONTENTS

PREFACE

Probability theory is the mathematical theory of random (nondeterministic) phenomena. This new book presents the latest research in the field.

Grade of membership (GoM) analysis was introduced in 1974 [27] as a means of analyzing multivariate categorical data. Since then, it has been successfully applied to many problems. The primary goal of GoM analysis is to derive properties of individuals based on results of multivariate measurements; such properties are given in the form of the expectations of a hidden random variable (state of an individual) conditional on the result of observations. In Chapter 1, we present a new perspective for the GoM model, based on considering the observed distribution law as a mixture of individual distribution laws, which are assumed to be independent. It happens that the mixing distribution is not fully identifiable from available data; nevertheless, it has a number of identifiable invariants, which are sufficient to derive substantial properties of the population under investigation. These identifiable invariants might be interpreted as hidden individual states in a low-dimensional phase space. An additional advantage of our approach is that it provides a basis for a new, very efficient numerical procedure for estimation of model parameters.

In Chapter 2, we discuss the strong regularity concept of Lorentz [8] applied to the convolution method of summability. Later, we obtain the summability functions and absolute summability functions of this method.

Semi-stable laws and max-semi-stable laws can be studied from two angles, (i) the semi-stability property of sums and maximums and (ii) their domains of partial attraction. In Chapter 3 we discuss these laws from both the angles in samples of random size and they are referred to as φ-semi-stable and φ-max-semi-stable laws, where φ is a Laplace transform. For a given Laplace transform φ, a class of probability generating functions is presented. The integer-valued random variables corresponding to these probability generating functions satisfy certain conditions in the transfer theorems for sums and maximums. This facilitates the discussion of limit theorems for φ-semi-stable, φ-max-semi-stable and operator φ-semi-stable laws. Then we study the discrete φ-semi-stable laws and their attraction. φ-stable and φ-max-stable laws are discussed generalizing certain results on geometric stable and geometric max-stable laws. We give a method to construct max-self-decomposable laws, semi-self-decomposable laws and max-semi-self-decomposable laws. In the process we also address the question; why certain integer valued random variables do not induce the semi-stability property in the corresponding random sums/ maximums. Notable integer valued random variables in this category are Poisson and the negative binomial on {0,1,2, ...}. A necessary and sufficient condition for this is given.

Conditional Equi-concentration of Types on I-projections is presented in Chapter 4. It provides an extension of Conditional Weak Law of Large Numbers to the case of several Iprojections. Also a multiple I-projections extension of Gibbs Conditioning Principle is developed. μ-projection variants of the probabilistic laws are stated. Implications of the results for Relative Entropy Maximization, Maximum Probability, Maximum Entropy in the Mean and Maximum R´enyi-Tsallis Entropy methods are discussed.

In Chapter 5 new product theorems of queuing theory have been proved. These theorems give accuracy analytical formulas to calculate stationary distributions in some queuing networks with complicated structures and protocols. Three directions of product theorems development have been considered in this paper. All these directions are closely connected with systems of mobile telephone connection [1]. First direction is product theorems in queuing networks with prohibitions. Such networks appear for example in queuing systems with retrial queues. A consideration of queuing network with prohibitions demands to decompose a system of the network motion equations in an accordance with a graph of permissible transitions. Second direction is a consideration of queuing network behavior in a random environment and its adaptation to this environment. For this aim a technique of conditional product stationary distributions is developed. These directions of product theorems development are closely connected with attendant transportation problems. Special algorithms of a search of permissible solutions of these problems have been constructed. Last direction is a consideration of queuing networks with switching centers in which a route of the customer doesn't change. Special properties of Poisson distribution allow to conserve product form of stationary distributions in queuing network with such centers.

One of the main goals of extreme value analysis is to estimate the probability of rare events given a sample from an unknown distribution. The upper tail behavior of this distribution is described by the extreme value index. In Chapter 6, we present a new estimator of the extreme value index adapted to any domain of attraction. Its construction is similar to the one of Pickands' estimator. Its weak consistency as well as its asymptotic distribution are established and a bias reduction method is proposed.

Chapter 7 presents, let Z_n, $n = 1, 2, \ . \ . \ .$ be the supercritical Galton-Watson process with the finite mean family size. Let f be a nonnegative, nondecreasing and concave on $[0, \infty)$ function. For $B, d \geq 0$ a new function g is defined by $g(x) := B + \int_d^x (f(y)/y)dy$ for $x \geq d$, and $g(x) = 0$ for $x < d$. Assume that, when $n \to \infty$, $Z_n/\mathsf{E}Z_n$ converges almost surely and in mean to a random variable W. By using a new (nonanalytic) approach it is proved that $\mathsf{E}Wf(W) < \infty$ if and only if $\mathsf{E}Z_1 g(Z_1) < \infty$.

The transport phenomena are known to be important for various scientific domains: examples can be found in physics, electrochemistry, heterogeneous catalysis, physiology, etc. To obtain new information about diffusive or Laplacian transport towards a semi-permeable or resistive interface, one can study the random trajectories of diffusive particles modelled, in a first approximation, by the partially reflected Brownian motion. This stochastic process turns out to be a convenient mathematical foundation for discrete, semi-continuous and continuous theoretical descriptions of the diffusive transport. Chapter 8 presents an overview on these topics with a special emphasis on the close relations between stochastic processes with partial reflections and the Laplacian transport phenomena. We give selected examples of these phenomena followed by a brief introduction to the partially reflected Brownian motion and related probabilistic objects (e.g., local time process and spread harmonic measure). A particular attention is paid to a useful relation to the Dirichlet-to-Neumann operator. Some practical consequences and further perspectives are discussed.

Chapter 9 studies occupation time on hypersurface and its variations for *n*-dimensional jump diffusion processes. It is shown that the occupation time on hypersurfaces does exists for the jump processes as a limit in variance for a wide class of piecewise smooth hypersurfaces, including some fractal type and moving surfaces. Solvability and uniqueness of integro-differential Kolmogorov–Fokker–Planck with generalized functions in coefficients are investigated. An analog of the Meyer-Tanaka formula is presented. For a process with varying drifts, diffusions, and jump parameters, formulas for variations of expectations of functionals of occupation time are derived, including variations of the characteristic functions.

In: Focus on Probability Theory
Editor: Louis R. Velle, pp. 1-26
ISBN 1-59454-474-3

Chapter 1

GRADE OF MEMBERSHIP ANALYSIS: ONE POSSIBLE APPROACH TO FOUNDATIONS*

Mikhail Kovtun † ***Igor Akushevich and Kenneth G. Manton***
Center for Demographic Studies, Duke University,
Durham, NC 27708
H. Dennis Tolley ‡
Department of Statistics, Brigham Young University,
Provo, UT 84602

Abstract

Grade of membership (GoM) analysis was introduced in 1974 [27] as a means of analyzing multivariate categorical data. Since then, it has been successfully applied to many problems. The primary goal of GoM analysis is to derive properties of individuals based on results of multivariate measurements; such properties are given in the form of the expectations of a hidden random variable (state of an individual) conditional on the result of observations.

In this article, we present a new perspective for the GoM model, based on considering the observed distribution law as a mixture of individual distribution laws, which are assumed to be independent.

It happens that the mixing distribution is not fully identifiable from available data; nevertheless, it has a number of identifiable invariants, which are sufficient to derive substantial properties of the population under investigation. These identifiable invariants might be interpreted as hidden individual states in a low-dimensional phase space.

An additional advantage of our approach is that it provides a basis for a new, very efficient numerical procedure for estimation of model parameters.

Key Words: Mixed distributions, latent structure analysis, grade of membership analysis, multivariate categorical data, linear regression, multidimensional distribution.

AMS Subject Classification: Primary 62H12; secondary 62J99.

---

*This research was supported by grants from National Institute of Aging.

†E-mail address: mkovtun@cds.duke.edu

‡E-mail address:tolley@byu.edu

1 Introduction

The grade of membership (GoM) analysis was initially introduced in [27]; the term "grade of membership" is due to this article. Since then, the GoM analysis has been successfully applied to many practical problems. The present work originated as an attempt to construct a foundation for the GoM analysis from the point of view of the theory of mixed distributions. The result appears to be a new class of models that differs from the original one both in methods of the model construction and in mathematical meaning of the estimators obtained from the model. We still prefer to keep the old name, "grade of membership analysis," as the application domain remains the same.

The class of GoM models belongs to a family of latent structure models, which, in turn, is a subfamily of a family of mixed distribution models.

Investigation of mixed distributions began near the end of 19th century, in works of Simon Newcomb and Karl Pearson [19, 21]. That time, special cases of a finite mixture of normal distributions were analysed. In the middle of the 20th century, a new wave of interest in mixtures began, primarily due to the works of Henry Teicher [24, 25, 26], who made fundamental definitions and introduced a crucial notion of identifiability.

In the simplest case of parametric family of distributions of real-valued random variables, mixtures may be defined as follows. Let $\mathcal{F} = \{F(x;\theta) \mid \theta \in \Theta\}$ be a family of cumulative distribution functions (c.d.f.), parametrized by a parameter θ; the set of possible values of parameter Θ is a subset of finite-dimensional real space $\mathbb{R}^n$. For example, $\mathcal{F}$ may be a family of normal distributions; in this case, $\theta = (m,\sigma)$, where m is expectation and σ is standard deviation; thus, Θ is the upper half-plane of the Euclidean plane. Let μ be a probability measure on Θ. Then a *mixture* of $\mathcal{F}$ defined by a *mixing distribution* μ is:

$$F_\mu(x) = \int_\Theta F(x;\theta)\mu(d\theta) \tag{1}$$

Let $\overline{\mathcal{F}}$ be a family of c.d.f. representable in form (1). The most important question regarding mixtures is the one of *identifiability*: for a given c.d.f. $F \in \overline{\mathcal{F}}$, whether the mixing distribution μ that produces F is unique?

It happens that mixtures of few families are identifiable. For example, a family of all normal distributions is not identifiable; only families of normal distributions with fixed mean or fixed variance are.

One way to achieve identifiability is to restrict the set of allowable measures μ in (1). If we allow only measures with finite support, we come to a notion of *finite mixtures*. In this case, (1) is replaced by

$$F_\mu(x) = \sum_{i=1}^{p} \mu_i F(x;\theta_i) \tag{2}$$

where $\{\theta_1,\ldots,\theta_p\}$ is a finite set supporting μ, and μ_i is a weight of point θ_i.

The main results regarding identifiability of mixtures are: (a) finite mixtures of $\mathcal{F}$ are identifiable if and only if $\mathcal{F}$ is linearly independent (i.e. for every $\theta_1,\ldots,\theta_p$, the equality $\sum_{i=1}^{p} \mu_i F(x;\theta_i) \equiv 0$ implies $\mu_i = 0$ for all i; (b) arbitrary mixtures of $\mathcal{F}$ are identifiable if and only if for arbitrary signed measure μ the equality $\int_\Theta F(x;\theta)\mu(d\theta) \equiv 0$ implies that μ is identically zero [26, 1, 20, 29, 22, 23].

Identifiability of finite mixtures is established for a wide range of families of distributions, including normal distributions, Poisson distributions, exponential families, etc. Still, many important from practical point of view families do not enjoy this property. For example, latent class models are not identifiable in general settings (they are identifiable only when the number of classes is sufficiently less than the number of measurements).

The mixed distribution models are natural choices when the population under question is assumed to be heterogeneous. The absence of identifiability, however, often prevents their usage. Another way to extend applicability of the mixed distribution models in the absence of identifiability is to search for their *invariants*, i.e. functions of mixing distributions that take the same values on all mixtures giving the observed distribution.

Some questions regarding mixture invariants were investigated by Bruce G. Lindsay [15, 16]. He has shown that some invariants may always be identified, regardless of identifiability of a mixture.

GoM models give an example of practically useful mixture models, in which the mixing distribution is not identifiable, but there exists a number of identifiable invariants, which, from a practical point of view, provide substantial knowledge about the population under consideration.

Latent structure models are a special kind of mixed distribution models. Latent structure analysis was introduced by Paul Lazarsfeld in 1950 [12, 13]; a more detailed exposition is given in [14, 3, 8, 11, 4, 10, 2].

Latent structure analysis considers J discrete measurements on each individual (like multi-choice questions), represented by random variables $X_1, \ldots, X_J$, with the set of outcomes of jth measurement being $\{1, \ldots, L_j\}$. It tries to represent the observed distribution law as a mixture of distribution laws, in which random variables $X_1, \ldots, X_J$ are independent. Depending on a family of allowed mixing distributions, different kinds of latent structure models emerge.

With this definition of the latent structure analysis, GoM models compose a subfamily of a family of latent structure models (and, consequently, of a family of mixed distribution models). This fact was not recognized, however, for a long time. Development of this approach, undertaken in the present article, lead to fruitful results. Still the technique used for construction of GoM models is very different from the one used in other branches of latent structure analysis.

From the point of view of applications, the goal of GoM analysis is to derive some properties of an individual based on results of measurements. Making a medical diagnosis is an example of such inference: a physician makes a decision about the health of an individual based on results of objective measurements and answers given by a patient, and the meaning of these results is derived from the results of similar measurements made on other individuals.

Mathematically, one possibility to express this is to assume that there exists a hidden continuous random variable G representing knowledge about individuals derivable from observations (in the diagnostic example, it is the health state of an individual.) Now one is interested in what might be said about value of G based on observed values of $X_1, \ldots, X_J$. More specifically, values of interest are expectations of G conditional on values of random variables $X_1, \ldots, X_J$, $\mathsf{E}(G \mid X_1 = x_1, \ldots, X_J = x_J)$.

The main result of the present article, contained in section 7, is that the values of interest (i.e. conditional expectations and conditional variances) are solution of system (38), and that under modest conditions, *only* values of interest are solutions of this system. Furthermore, as corollary 7.6 shows, the system (38) can be solved by a two-step process, each step of which consists of solving problem of linear algebra.

GoM analysis (as well as many flavors of latent structure analysis) employs an assumption that the problem under consideration has a lower dimensionality than observed data. Our theorem 7.3 and its corollary gives a way to estimate this dimensionality directly (which usually presents a substantial problem is such kind of analysis.)

An additional advantage of our approach is that it not only establishes a way to estimate values of interest, but also provides a ground for evaluation of confidence intervals (not addressed in the present article.)

The rest of the article is organized as follows:

In section 3, we mathematically formulate the problem and define related notions. The central idea here is to consider the observed distribution law as a mixture of individual distribution laws. We show that initial data are sufficient to estimate a set of moments of the mixing distribution up to order J (the number of measurements.)

In section 4, we consider the GoM problem as a problem of finding a low-dimensional distribution and obtain basic corollaries of this hypothesis.

In section 5, we consider a hypothesis that there exists a linear regression of observed random variables X_j on hidden random variable G. We show that this hypothesis is essentially equivalent to the one considered in previous section.

In section 6, we establish relations between distributions and moments of β and G, and find transformation laws for changing their basis. The main result of this section is equation (31).

In section 7, we consider a system of equations (38). We show that values of interest are always solutions of this system, and we establish sufficient conditions under which the system (38) has *only* such solutions.

In section 8, we outline a numerical procedure for estimating values of interest and discuss its properties.

2 Preliminaries

2.1 Notation

$\mathbb{Z}$ is the set of integers, and $\mathbb{R}$ is the set of reals. $\mathbb{Z}^+$ and $\mathbb{R}^+$ are subsets of positive, and $\mathbb{Z}^{+0}$ and $\mathbb{R}^{+0}$ are subsets of nonnegative, integers and reals, respectively.

For $m,n \in \mathbb{Z}$, $[m..n]$ denotes the set of integers between m and n: $[m..n] = \{z \in \mathbb{Z} \mid m \leq z \leq n\}$. If $m > n$, $[m..n] = \varnothing$.

$\mathbb{R}^n$ is n-dimensional linear space over reals, and $\mathbb{S}^n$ is a $(n-1)$-dimensional unit simplex in $\mathbb{R}^n$, $\mathbb{S}^n = \{x \in \mathbb{R}^n \mid x_i \geq 0 \text{ and } \sum_i x_i = 1\}$.

For a linear subspace $Q \subseteq \mathbb{R}^n$, $\dim(Q)$ denotes its dimension.

For $x^1,\ldots,x^p \in \mathbb{R}^n$, $\mathrm{Lin}(x^1,\ldots,x^p)$ denotes a linear subspace of $\mathbb{R}^n$ spanned by vectors $x^1,\ldots,x^p$, and $\mathrm{rank}(x^1,\ldots,x^p)$ denotes a rank of system of vectors $x^1,\ldots,x^p$ (thus, $\mathrm{rank}(x^1,\ldots,x^p) = \dim(\mathrm{Lin}(x^1,\ldots,x^p))$.)

For $a \in \mathbb{R}$ (or $a \in \mathbb{Z}$) and $i \in [1..n]$, $\boldsymbol{a}_i$ (with boldface a) denotes a vector from $\mathbb{R}^n$ ($\mathbb{Z}^n$, respectively) with i^{th} component equal a and all other components equal 0. Dimensionality of $\boldsymbol{a}_i$ will be clear from context.

2.2 Support of Measures

We consider only probability measures defined on σ-algebra of Borel sets of $\mathbb{R}^n$; a measure μ is a probability measure, if $\mu(\mathbb{R}^n) = 1$.

A *support* of measure μ is a closed set $A \subseteq \mathbb{R}^n$ such that $\mu(A) = 1$. We do not require a minimality of a support: if A is a support of μ and $A \subseteq A'$, A' is closed, then A' also is a support of μ.

We use $\mathrm{Supp}(\mu)$ to denote the set of all supports of μ. Thus, $A \in \mathrm{Supp}(\mu)$ means "A is a support of μ." Note that $A \in \mathrm{Supp}(\mu)$ implies that A is closed.

2.3 Indexing Contingency Tables and Related Objects

We need a way for indexing cells in a contingency table and for other objects having similar structure.

A contingency table for a set of J discrete measurements, with L_j possible outcomes for measurement j, is a J-dimensional table having $L_j + 1$ cells in dimension j. Index for j^{th} dimension ranges from 0 to L_j.

More formally, let $\mathfrak{L}_\infty = \{(l_1, \ldots, l_J) \mid l_j \in \mathbb{Z}^+\}$ and $\mathfrak{L}^0_\infty = \{(l_1, \ldots, l_J) \mid l_j \in \mathbb{Z}^{+0}\}$, i.e. sets of J-dimensional vectors with positive and, respectively, nonnegative integer components. There is a one-to-one correspondence between sets of J discrete measurements and vectors in $\mathfrak{L}_\infty$: a vector $L = (L_1, \ldots, L_J)$ describes a set of J measurements, in which measurement j has L_j outcomes.

For every $L \in \mathfrak{L}_\infty$, let $\mathfrak{L}_L = \{\ell \in \mathfrak{L}_\infty \mid \ell_j \leq L_j\}$ and $\mathfrak{L}^0_L = \{\ell \in \mathfrak{L}^0_\infty \mid \ell_j \leq L_j\}$. If L defines a set of measurements, $\mathfrak{L}_L$ is a set of all possible outcomes of these measurements, and $\mathfrak{L}^0_L$ is a complete set of indices for the contingency table. In addition, for every $\mathfrak{J} \subseteq [1..J]$, let $\mathfrak{L}^{[\mathfrak{J}]}_L = \{\ell \in \mathfrak{L}^0_L \mid \ell_j = 0 \Leftrightarrow j \in \mathfrak{J}\}$. The set $\mathfrak{J}$ indicates measurements that we exclude from consideration, and vector $\ell \in \mathfrak{L}^{[\mathfrak{J}]}_L$ contains results of all measurements except those listed in $\mathfrak{J}$. Note that $\mathfrak{L}^{[\varnothing]}_L = \mathfrak{L}_L$ and $\mathfrak{L}^0_L = \cup_{\mathfrak{J} \subseteq [1..J]} \mathfrak{L}^{[\mathfrak{J}]}_L$.

Vector $\ell' \in \mathfrak{L}^{[\mathfrak{J}]}_L$ may be considered as describing a family of outcomes $\{\ell \in \mathfrak{L}_L \mid \ell_j = \ell'_j \text{ for } j \notin \mathfrak{J}\}$. Abusing notation, we will also use ℓ' to denote this family. More generally, we write $\ell' \in \ell''$ for $\ell' \in \mathfrak{L}^{[\mathfrak{J}']}_L$ and $\ell'' \in \mathfrak{L}^{[\mathfrak{J}'']}_L$ whenever $\ell'_j = \ell''_j$ for all $j \notin \mathfrak{J}''$ (note $\ell' \in \ell''$ is possible only when $\mathfrak{J}' \subseteq \mathfrak{J}''$.) For $\ell \in \mathfrak{L}^{[\mathfrak{J}]}_L$, let $\ell^{[\mathfrak{J}']}$ be a vector from $\mathfrak{L}^{[\mathfrak{J} \cup \mathfrak{J}']}_L$ such that $\ell^{[\mathfrak{J}']}_j = \ell_j$ for all $j \notin \mathfrak{J}'$. We always have $\ell \in \ell^{[\mathfrak{J}']}$. We write $\ell^{[j]}$, $\ell^{[j_1, j_2]}$, etc. instead of $\ell^{[\{j\}]}$, $\ell^{[\{j_1, j_2\}]}$, etc., respectively.

Let also set $|L| = \sum_j L_j$ and $|L^*| = \prod_j L_j$.

We always assume that the set of our measurements is described by a vector L. We drop index L in notations $\mathfrak{L}_L$ and $\mathfrak{L}^{[\mathfrak{J}]}_L$ if it does not create an ambiguity.

A contingency table may be constructed for any sample by putting in the cell with index ℓ the number of individuals who (a) have outcome ℓ_j for measurement j if $\ell_j \neq 0$; and (b) have arbitrary outcomes for all other measurements. Let N_ℓ be a value in ℓ^{th} cell of contingency table. The usual summation rule for the contingency tables in our notation is: for any $\mathfrak{J}' \subseteq \mathfrak{J} \subseteq [1..J]$, $\ell \in \mathfrak{J}$, $N_\ell = \sum_{\ell' \in \mathfrak{L}^{[\mathfrak{J}']} : \ell' \in \ell} N_{\ell'}$. Note that $N = N_{(0,\ldots,0)}$ is the sample size.

A frequency table is obtained from a contingency table by dividing the value in each cell by N. We use f_ℓ to denote a value of ℓth cell of a frequency table. The above summation rule is applicable to frequency tables as well.

3 The Problem

We consider a population of a potentially infinite number of individuals, every individual being subject to J measurements with discrete outcomes. Without loss of generality, we may assume that outcomes of jth measurement are $\{1,\ldots,L_j\}$.

The outcomes of these measurements are represented by a random vector $X = (X_1,\ldots, X_J)$, which takes values in $\mathfrak{L}_L$. The distribution law of this random vector is given by an $|L^*|$-dimensional vector of probabilities $\widetilde{\beta} = (\widetilde{\beta}_\ell)_{\ell\in\mathfrak{L}_L}$, where $\widetilde{\beta}_\ell = \mathsf{P}(X_1 = \ell_1 \wedge\cdots\wedge X_J = \ell_J)$.

We say that the distribution law of the random vector X is *independent*, if component random variables are mutually independent, i.e. if for every $\ell \in \mathfrak{L}^0$

$$\mathsf{P}\left(\bigwedge_{j:\ell_j\neq 0} X_j = \ell_j\right) = \prod_{j:\ell_j\neq 0} \mathsf{P}(X_j = \ell_j) \tag{3}$$

The first GoM assumption (which is a standard assumption of the latent structure analysis) is:

(G1) The observed distribution law $\widetilde{\beta}^0$ is a mixture of independent distribution laws, i.e.

$$\widetilde{\beta}^0 = \int \widetilde{\beta}\mu_{\widetilde{\beta}}(d\widetilde{\beta}) \tag{4}$$

where $\mu_{\widetilde{\beta}}$ is supported by a family of independent distribution laws.

An independent distribution law is fully described by an $|L|$-dimensional vector of probabilities $\beta = (\beta_{jl})_{jl}$ ($j \in [1..J]$, and for every j, $l \in [1..L_j]$), where $\beta_{jl} = \mathsf{P}(X_j = l)$. As $\mu_{\widetilde{\beta}}$ is supported by the set of independent distribution laws, there exists the equivalent measure μ_β on $\mathbb{R}^{|L|}$. With this, (4) may be rewritten as:

$$\forall \ell \in \mathfrak{L} : \quad \widetilde{\beta}^0_\ell = \int \left(\prod_{j=1}^{J} \beta_{j\ell_j}\right)\mu_\beta(d\beta) \tag{5}$$

We start with elementary properties, which may be directly derived from definitions.

As β_{jl} are probabilities, they satisfy

$$\text{(a)} \quad \beta_{jl} \geq 0 \qquad\qquad \text{(b)} \quad \text{for all } j\text{:} \quad \sum_{l=1}^{L_j} \beta_{jl} = 1 \tag{6}$$

Thus, a product of simplices $\mathbb{S}^L = \prod_j \mathbb{S}^{L_j} \subseteq \mathbb{R}^{|L|}$ is a support of the measure μ_β, $\mathbb{S}^L \in \text{Supp}(\mu_\beta)$.

The equation (5) may be generalized to:

$$\forall \ell \in \mathfrak{L}^0 : \quad \mathsf{P}\left(\bigwedge_{j:\ell_j\neq 0} X_j = \ell_j\right) = \int \left(\prod_{j:\ell_j\neq 0} \beta_{j\ell_j}\right)\mu_\beta(d\beta) \tag{7}$$

For every $\ell \in \mathcal{L}^0$, let the ℓ-moment of distribution μ_β be

$$M_\ell(\mu_\beta) = \int \left(\prod_{j:\ell_j \neq 0} \beta_{j\ell_j} \right) \mu_\beta(d\beta) \tag{8}$$

In particular, we have $\mathcal{L}^{[1,\ldots,J]} = \{(0,\ldots,0)\}$, and $M_{(0,\ldots,0)}(\mu_\beta) = \int \mu_\beta(d\beta) = 1$.

Comparing (8) with (7), we see that the ℓ-moment of distribution μ_β is equal to the probability of set of outcomes ℓ.

For $\ell \in \mathcal{L}^{[\mathcal{J}]}$, $M_\ell(\mu_\beta)$ is $J - |\mathcal{J}|$ order mixed moment of μ_β. The set of ℓ-moments for all $\ell \in \mathcal{L}^0$ does not exhaust, however, the set of all moments of order up to J (for example, a moment $\int \beta_{11}\beta_{12}\mu_\beta(d\beta)$ is not an ℓ-moment.) At the end of section 7 we shall discuss in more detail whether $\{M_\ell(\mu_\beta)\}_\ell$ can determine all moments of order up to J.

A basic statistical fact is that frequencies f_ℓ are consistent and efficient estimators for $M_\ell(\mu_\beta)$.

The following proposition and its corollary is an equivalent of the summation rule for contingency and frequency tables.

Proposition 3.1 *Let $\mathcal{J}' \subseteq \mathcal{J}'' \subseteq [1..J]$. Then for every $\ell'' \in \mathcal{L}^{[\mathcal{J}'']}$*

$$M_{\ell''}(\mu_\beta) = \sum_{\ell' \in \mathcal{L}^{[\mathcal{J}']}\,:\,\ell' \in \ell''} M_{\ell'}(\mu_\beta)$$

Proof. For every $j_0 \in \mathcal{J}'' \setminus \mathcal{J}'$ and for every $\ell' \in \mathcal{L}^{[\mathcal{J}']}$ we have:

$$\sum_{\ell \in \mathcal{L}^{[\mathcal{J}']}\,:\,\ell \in \ell'^{\cup j_0}} M_\ell(\mu_\beta) = \sum_{\ell \in \mathcal{L}^{[\mathcal{J}']}\,:\,\ell \in \ell'^{\cup j_0}} \int \prod_{j \notin \mathcal{J}'} \beta_{j\ell_j}\mu_\beta(d\beta) =$$

$$\sum_{\ell \in \mathcal{L}^{[\mathcal{J}']}\,:\,\ell \in \ell'^{\cup j_0}} \int \beta_{j_0\ell_{j_0}} \cdot \prod_{j \notin \mathcal{J}' \cup \{j_0\}} \beta_{j\ell_j}\mu_\beta(d\beta) = \sum_{l=1}^{L_{j_0}} \int \beta_{j_0 l} \cdot \prod_{j \notin \mathcal{J}' \cup \{j_0\}} \beta_{j\ell_j}\mu_\beta(d\beta) =$$

$$\int \left(\sum_{l=1}^{L_{j_0}} \beta_{j_0 l} \right) \cdot \prod_{j \notin \mathcal{J}' \cup \{j_0\}} \beta_{j\ell_j}\mu_\beta(d\beta) = \int 1 \cdot \prod_{j \notin \mathcal{J}' \cup \{j_0\}} \beta_{j\ell_j}\mu_\beta(d\beta) = M_{\ell'^{\cup j_0}}(\mu_\beta)$$

The rest of the proof is induction over the size of $\mathcal{J}'' \setminus \mathcal{J}'$. ■

Corollary 3.2 *For every $\mathcal{J} \subseteq [1..J]$, $\sum_{\ell \in \mathcal{L}^{[\mathcal{J}]}} M_\ell(\mu_\beta) = 1$. In particular, $\sum_{\ell \in \mathcal{L}} M_\ell(\mu_\beta) = 1$.*

Below we consider another two (essentially equivalent) assumptions. The first one is that a support of μ_β is restricted to $(K-1)$-dimensional affine plane in $\mathbb{R}^{|L|}$. The second assumption is that there exists a random variable G taking values in $\mathbb{R}^K$ such that there exists a linear regression of random variables $X_1,\ldots,X_J$ on G.

4 Low-dimensional Distributions

The second assumption that we consider is:

(G2′) The support of μ_β is a K-dimensional linear subspace Q of $\mathbb{R}^{|L|}$, and any proper subspace of Q does not support μ_β.

We include the second clause (no proper subspace of Q supports μ_β) to avoid degenerate cases. Any degenerate case may be considered as a nondegenerate case for some $K' < K$.

As $\mathbb{S}^L \in \mathrm{Supp}(\mu_\beta)$, the intersection $P_\beta = Q \cap \mathbb{S}^L$ is necessarily nonempty, and this intersection supports μ_β. In general, P_β is $(K-1)$-dimensional polyhedral body, which has at least K vertices. Let $\bar{P}_\beta$ be the $(K-1)$-dimensional affine space spanned by P_β.

Let $\Lambda = \{\lambda^1, \ldots, \lambda^K\}$ be a linear basis of Q. We also consider Λ as a $|L| \times K$ matrix,

$$\Lambda = \begin{pmatrix} \lambda^1_{11} & \cdots & \lambda^K_{11} \\ & \vdots & \\ \lambda^1_{JL_J} & \cdots & \lambda^K_{JL_J} \end{pmatrix} \tag{9}$$

There exists considerable freedom in choosing Λ. We shall exploit it by imposing constraints on Λ. The first one is:

(Λ_0) For every k, $\lambda^k \in \bar{P}_\beta$.

Let $g = (g_1, \ldots, g_K)$ be a vector of coordinates of a point $\beta \in Q$ in basis Λ, i.e. $\beta = \sum_{k=1}^K g_k \lambda^k$. Then ($\Lambda_0$) implies

$$\sum_{k=1}^K g_k \lambda^k \in \bar{P}_\beta \quad \Leftrightarrow \quad \sum_{k=1}^K g_k = 1 \tag{10}$$

If Λ and Λ' are two bases of Q, there exists a nondegenerate $K \times K$ matrix $A = (a_k^{k'})_{k'k}$ such that $\Lambda' = \Lambda A$.

Using the fact that $(1, \ldots, 1)$ is left eigenvector of matrix A corresponding to eigenvalue 1 if and only if every column of A sums to 1, $\sum_k a_k^{k'} = 1$, one easily obtains the following two propositions:

Proposition 4.1 *Let both Λ and Λ' satisfy (Λ_0). Then $(1, \ldots, 1)$ is left eigenvector of matrix A with eigenvalue* 1.

Proposition 4.2 *Let Λ satisfy (Λ_0) and let A be a nonsingular matrix with left eigenvector $(1, \ldots, 1)$ with eigenvalue* 1. *Then $\Lambda' = \Lambda A$ satisfies (Λ_0).*

If g is a coordinate vector of $\beta \in Q$ in basis Λ, $\beta = \Lambda g$, then the coordinate vector of β in basis $\Lambda' = \Lambda A$ is $g' = A^{-1} g$.

Remark 4.3 In matrix expressions (like $\beta = \Lambda g$ above,) we always assume that all vectors are columns. ■

Every choice of a basis Λ induces a linear map:

$$H_\Lambda : \mathbb{R}^K \to Q, \qquad H_\Lambda(g) = \sum_k g_k \lambda^k \tag{11}$$

Note that Λ is a matrix of linear map H_Λ with respect to basis Λ in Q and standard unit basis in $\mathbb{R}^K$.

When the basis Λ satisfies (Λ_0), $H_\Lambda^{-1}(\bar{P}_\beta)$ is a unit affine plane $\bar{P}_g$ in $\mathbb{R}^K$, $\bar{P}_g = \{g \in \mathbb{R}^K \mid \sum_k g_k = 1\}$, and $P_g^\Lambda = H_\Lambda^{-1}(P_\beta)$ is a convex $(K-1)$-dimensional polyhedron in $\bar{P}_g$.

The map H_Λ allows us to introduce a measure μ_g^Λ on $\bar{P}_g$, defined as:

$$\mu_g^\Lambda(B) = \mu_\beta(H_\Lambda(B)) \qquad \text{for every Borel set } B \subseteq \bar{P}_g \tag{12}$$

As $P_\beta \in \mathrm{Supp}(\mu_\beta)$, we have $P_g^\Lambda \in \mathrm{Supp}(\mu_g^\Lambda)$.

Thus, we can replace integration over P_β by integration over P_g^Λ:

$$\int_{P_\beta} \phi(\beta)\mu_\beta(d\beta) = \int_{P_g^\Lambda} \phi(H_\Lambda(g))\mu_g^\Lambda(dg) \tag{13}$$

for every measurable function ϕ.

Remark 4.4 We are trying to reflect in our notation all substantial dependencies between objects. Measure μ_β and polyhedron P_β, of course, do not depend on the choice of Λ; thus, no index Λ in notation μ_β and P_β. On the contrary, map H defined by (11) (and consequently polyhedron P_g and measure μ_g defined by (12)) substantially depends on the choice of Λ — so we use notation H_Λ, P_g^Λ, and μ_g^Λ. However, we shall drop the index Λ in the above notation if it is obvious from the context. ■

5 Linear Regression Hypothesis

A random variable X_j has a finite range $[1..L_j]$, on which no arithmetic operations are defined. This prevents us from considering expectation, variance, etc. of X_j. To cope with this problem, we associate with every X_j a random vector Y_j taking values in $\mathbb{R}^{L_j}$ and defined as: if $X_j = l$, then $Y_j = \mathbf{1}_l$, (recall that $\mathbf{1}_l$ is a L_j-dimensional vector with l^{th} component equals 1, and all other components equal 0).

There is an important connection between distributions of X_j and Y_j: if $(\beta_{jl})_l$ is a vector of probabilities of X_j, $\beta_{jl} = \mathsf{P}(X_j = l)$, then $\mathsf{E}(Y_j) = (\beta_{jl})_l$ (here and below $\mathsf{E}(\cdot)$ denotes expectation.) In general, for every condition C we have $\mathsf{E}(Y_j \mid C) = (\mathsf{P}(X_j = l \mid C))_l$.

Remark 5.1 As Y_j is an L_j-dimensional vector, $\mathsf{E}(Y_j)$ is also an L_j-dimensional vector. We use $\mathsf{E}_m(\cdot)$ to denote m^{th} component of vector expectation. ■

Thus, we have

Proposition 5.2 *For every j and for every condition C,* (a) $\mathsf{E}_l(Y_j \mid C) \geq 0$, *and* (b) $\sum_l \mathsf{E}_l(Y_j \mid C) = 1$.

Now we can formulate an alternative form of assumption (G2):

(G2″) There exists a random vector G, which takes values in $\mathbb{R}^K$, such that:

(a) There exists a joint distribution of G and X.

(b) Conditional random variables $(X_1 \mid G=g), \ldots, (X_J \mid G=g)$ are mutually independent for every g.

(c) For every j, a regression of Y_j on G is linear.

(d) For any $K' < K$ there is no random vector G' satisfying (a)–(c).

Again, clause (d) is intended to prevent degenerate cases.

Clause (c) means that for every j, there exist vectors $(\lambda^1_{jl})_l, \ldots, (\lambda^K_{jl})_l$ such that

$$\mathsf{E}(Y_j \mid G=g) = \Big(\sum_k g_k \lambda^k_{jl}\Big)_l \tag{14}$$

or, in matrix form,

$$\mathsf{E}(Y_j \mid G=g) = \begin{pmatrix} \lambda^1_{j1} & \cdots & \lambda^K_{j1} \\ & \vdots & \\ \lambda^1_{jL_j} & \cdots & \lambda^K_{jL_j} \end{pmatrix} \cdot \begin{pmatrix} g_1 \\ \vdots \\ g_K \end{pmatrix} = \Lambda_j \cdot g \tag{15}$$

Taking into account the relation between $\mathsf{E}(Y_j)$ and probability distribution of X_j, one obtains:

Theorem 5.3 *(G2′) holds if, and only if, (G2″) holds.*

The random vector G, if it exists, is not defined uniquely: for every nondegenerate $K \times K$ matrix A, random vector $G' = A^{-1}G$ also satisfies (G2″), as:

$$\mathsf{E}(Y_j \mid G'=g') = \mathsf{E}(Y_j \mid AG'=Ag') = \mathsf{E}(Y_j \mid G=Ag') = \\ \Lambda_j \cdot (A \cdot g') = (\Lambda_j \cdot A) \cdot g' = \Lambda'_j \cdot g' \tag{16}$$

This nonuniqueness corresponds to the nonuniqueness of the basis for Q discussed in section 4. Again, one may choose G in such a way that (Λ_0) is satisfied.

Corollary 5.4 *In the presence of* (Λ_0)*, the possible values of* G *satisfy* $\sum_k g_k = 1$. *In other words,* G *takes values in a unit affine plane* $\bar{P}_g \subseteq \mathbb{R}^K$.

Corollary 5.5 *In the presence of* (Λ_0)*, a set of possible values of* G *is a bounded polyhedron* $P_g \subseteq \bar{P}_g$.

We are primarily interested in what can be said about value of G given outcomes of $X_1, \ldots, X_J$. The most interesting values are $\mathsf{E}(G \mid X = \ell)$ and $\mathsf{D}(G \mid X = \ell)$ (were $\mathsf{D}(\cdot)$ denotes variance.) We shall derive equations for these values in the next section.

6 Relations between μ_β and μ_g

As (G2′) and (G2″) are equivalent, we refer to (either of) them as (G2).

Under condition (G2) we have two distributions, μ_β and μ_g^Λ, connected by (11) and (12). In this section we establish further relations between μ_β and μ_g^Λ.

Throughout this section, we assume that some basis Λ of Q is fixed. We drop index Λ in all notation; however, the reader has to keep in mind that distribution μ_g, as well as all its moments, depend on Λ.

6.1 Unconditional Moments

We can express ℓ-moments of μ_β via moments of μ_g. Let $\mathcal{J} \subseteq [1..J]$ and $\ell \in \mathcal{L}^{[\mathcal{J}]}$. Then:

$$M_\ell(\mu_\beta) = \int_{P_\beta} \left(\prod_{j \notin \mathcal{J}} \beta_{j\ell_j} \right) \mu_\beta(d\beta) = \int_{P_g} \left(\prod_{j \notin \mathcal{J}} \sum_k g_k \lambda^k_{j\ell_j} \right) \mu_g(dg) =$$

$$\int_{P_g} \left(\sum_{w \in \mathcal{W}^{[\mathcal{J}]}} \left(\prod_{j \notin \mathcal{J}} g_{w_j} \cdot \prod_{j \notin \mathcal{J}} \lambda^{w_j}_{j\ell_j} \right) \right) \mu_g(dg) = \sum_{w \in \mathcal{W}^{[\mathcal{J}]}} \left(\left(\int_{P_g} \prod_{j \notin \mathcal{J}} g_{w_j} \mu_g(dg) \right) \cdot \prod_{j \notin \mathcal{J}} \lambda^{w_j}_{j\ell_j} \right) =$$

$$\sum_{w \in \mathcal{W}^{[\mathcal{J}]}} \left(M_w(\mu_g) \cdot \prod_{j \notin \mathcal{J}} \lambda^{w_j}_{j\ell_j} \right) \tag{17}$$

Here $\mathcal{W}^{[\mathcal{J}]} = \{(w_1, \ldots, w_J) \mid w_j \in [1..K] \text{ if } j \notin \mathcal{J},\ w_j = 0 \text{ if } j \in \mathcal{J}\}$, and for $w \in \mathcal{W}^{[\mathcal{J}]}$,

$$M_w(\mu_g) = \int_{P_g} \left(\prod_{j : w_j \neq 0} g_{w_j} \right) \mu_g(dg) \tag{18}$$

is a $(J - |\mathcal{J}|)^{\text{th}}$ order mixed moment of measure μ_g^Λ.

Note that $\mathcal{W}^{[\mathcal{J}]} = \mathcal{L}^{[\mathcal{J}]}_{(K,\ldots,K)}$. Thus, we freely apply to $\mathcal{W}$ all notations and conventions developed for $\mathcal{L}$ in section 2.3.

The sets of indices $\mathcal{W}^{[\mathcal{J}]}$ are redundant in the sense that different elements of $\mathcal{W}^{[\mathcal{J}]}$ correspond to the same moments. However, $\mathcal{W}^{[\mathcal{J}]}$ has the following nice property:

Proposition 6.1 *Let $\mathcal{J}' \subseteq \mathcal{J}'' \subseteq [1..J]$. Then for every $w'' \in \mathcal{W}^{[\mathcal{J}'']}$*

$$M_{w''}(\mu_g) = \sum_{w' \in \mathcal{W}^{[\mathcal{J}']} \,:\, w' \in w''} M_{w'}(\mu_g)$$

Proof. Similar to the proof of proposition 3.1. ■

Corollary 6.2 *For every $\mathcal{J} \subseteq [1..J]$, $\sum_{w \in \mathcal{W}^{[\mathcal{J}]}} M_w(\mu_g) = 1$. In particular, $\sum_{w \in \mathcal{W}} M_w(\mu_g) = 1$.*

To handle redundancy of $\mathcal{W}$, we introduce a new set of indices.

Let $\mathcal{V}[J', K'] = \{(v_1, \ldots, v_{K'}) \mid v_k \in [0..J'] \text{ and } \sum_k v_k = J'\}$. We write $\mathcal{V}[J']$ instead of $\mathcal{V}[J', K]$ and $\mathcal{V}$ instead of $\mathcal{V}[J, K]$.

For every $\mathcal{J} \subseteq [1..J]$ and for every $v \in \mathcal{V}[J - |\mathcal{J}|]$, let $\mathcal{W}_v^{[\mathcal{J}]} = \{w \in \mathcal{W}^{[\mathcal{J}]} \mid$ for every k, w contains exactly v_k components equal $k\}$. Let also $C_v^{[\mathcal{J}]} = |\mathcal{W}_v^{[\mathcal{J}]}|$.

Proposition 6.3

$$\text{(a)}\quad |\mathcal{V}[J',K']| = \frac{(J'+K'-1)!}{J'!(K'-1)!} \qquad\qquad \text{(b)}\quad C_v^{[\mathfrak{J}]} = \frac{(J-|\mathfrak{J}|)!}{v_1!\dots v_K!}$$

Proof. (a) By induction over $J'+K'$ from a recurrent equality $|\mathcal{V}[J',K']| = |\mathcal{V}[J'-1,K']| + |\mathcal{V}[J',K'-1]|$.

(b) Let $J' = J - |\mathfrak{J}|$. By direct computation one obtains:

$$C_v^{[\mathfrak{J}]} = \binom{J'}{v_1}\cdot\binom{J'-v_1}{v_2}\dots\binom{J'-\sum_{k=1}^{K-1}v_k}{v_K}$$

from which the statement of the proposition is straightforward. ■

One corollary to proposition 6.3 is that $C_v^{[\mathfrak{J}]}$ depends on $\mathfrak{J}$ only through $|\mathfrak{J}|$, and this value is contained in index v; thus, we can safely drop index $[\mathfrak{J}]$ and write just C_v.

As for every $w,w' \in \mathcal{W}_v^{[\mathfrak{J}]}$ we have $M_w(\mu_g) = M_{w'}(\mu_g)$ for every measure μ_g, we can define *v-moments of a measure* μ_g as

$$M_v(\mu_g) = M_{w_0}(\mu_g) = \int \prod_k g_k^{v_k}\mu_g(dg) \tag{19}$$

and *normalized v-moments of a measure* μ_g as

$$\bar{M}_v(\mu_g) = \sum_{w\in\mathcal{W}_v^{[\mathfrak{J}]}} M_w(\mu_g) = C_v M_{w_0}(\mu_g) = C_v\int\prod_k g_k^{v_k}\mu_g(dg) = C_v M_v(\mu_g) \tag{20}$$

In both equations, w_0 is an arbitrary element of $\mathcal{W}_v$. Note that both $M_v(\mu_g)$ and $\bar{M}_v(\mu_g)$ do not depend on $\mathfrak{J}$.

$\mathcal{V}[J']$ is the smallest possible set of indices for J'-order mixed moments of μ_g. Multiplier C_v in (20) allows us to obtain

Proposition 6.4 *For every* μ_g,

$$\sum_{v\in\mathcal{V}[J']}\bar{M}_v(\mu_g) = \sum_{v\in\mathcal{V}[J']} C_v M_v(\mu_g) = 1$$

Proof. Follows from proposition 6.1. ■

Now we can continue (17):

$$\sum_{w\in\mathcal{W}^{[\mathfrak{J}]}}\left(M_w(\mu_g)\cdot\prod_{j\notin\mathfrak{J}}\lambda_{j\ell_j}^{w_j}\right) = \sum_{v\in\mathcal{V}[J']}\sum_{w\in\mathcal{W}_v^{[\mathfrak{J}]}}\left(M_w(\mu_g)\cdot\prod_{j\notin\mathfrak{J}}\lambda_{j\ell_j}^{w_j}\right) =$$
$$\sum_{v\in\mathcal{V}[J']}\left(M_v(\mu_g)\cdot\sum_{w\in\mathcal{W}_v^{[\mathfrak{J}]}}\prod_{j\notin\mathfrak{J}}\lambda_{j\ell_j}^{w_j}\right) = \sum_{v\in\mathcal{V}[J']}\left(M_v(\mu_g)\cdot\sum_{w\in\mathcal{W}_v^{[\mathfrak{J}]}}\prod_{j\notin\mathfrak{J}}\lambda_{j\ell_j}^{w_j}\right) =$$
$$\sum_{v\in\mathcal{V}[J']}\left(M_v(\mu_g^{\Lambda})\cdot\Lambda(\mathfrak{J},v,\ell)\right) \tag{21}$$

where $J' = J - |\mathfrak{J}|$ and

$$\Lambda(\mathfrak{J},v,\ell) = \sum_{w\in\mathcal{W}_v^{[\mathfrak{J}]}}\prod_{j\notin\mathfrak{J}}\lambda_{j\ell_j}^{w_j} \tag{22}$$

6.2 Conditional Moments

For the joint distribution of $X = (X_1, \dots, X_J)$ and G we have, on the one hand,

$$d\,\mathsf{P}(G=g \wedge X=\ell) = \mathsf{P}(X=\ell \mid G=g)\cdot d\,\mathsf{P}(G=g) = \Big(\prod_j \sum_k g_k \lambda^k_{j\ell_j}\Big)\mu_g(dg) \tag{23}$$

and, on the other hand,

$$d\,\mathsf{P}(G=g \wedge X=\ell) = d\,\mathsf{P}(G=g \mid X=\ell)\cdot \mathsf{P}(X=\ell) = d\,\mathsf{P}(G=g \mid X=\ell)\cdot M_\ell(\mu_\beta) \tag{24}$$

Combining (23) and (24), one obtains

$$d\,\mathsf{P}(G=g \mid X=\ell) = \frac{\prod_j \sum_k g_k \lambda^k_{j\ell_j}}{M_\ell(\mu_\beta)}\mu_g(dg) \tag{25}$$

Similarly, for every $\mathcal{J} \subseteq [1..J]$ and for every $\ell \in \mathcal{L}^{[\mathcal{J}]}$,

$$d\,\mathsf{P}(G=g \mid X=\ell) = \frac{\prod_{j\notin\mathcal{J}} \sum_k g_k \lambda^k_{j\ell_j}}{M_\ell(\mu_\beta)}\mu_g(dg) \tag{26}$$

where for $\ell \in \mathcal{L}^{[\mathcal{J}]}$, $X=\ell$ means $\bigwedge_{j\notin\mathcal{J}} X_j = \ell_j$.

This allows us to conclude that the conditional distribution of $G \mid X=\ell$ is absolutely continuous with respect to measure μ_g, and

$$p_\ell(g) = \frac{\prod_{j\notin\mathcal{J}} \sum_k g_k \lambda^k_{j\ell_j}}{M_\ell(\mu_\beta)} \tag{27}$$

is its probability density function.

Having this, we may write (for every J', $v \in \mathcal{V}[J']$, $\mathcal{J} \subseteq [1..J]$, $\ell \in \mathcal{L}^{[\mathcal{J}]}$) a v-order mixed moment of G conditional on $X=\ell$

$$\mathsf{E}(G^v \mid X=\ell) = \int g^v p_\ell(g)\mu_g(dg) \tag{28}$$

where g^v denotes $\prod_k g_k^{v_k}$.

A special case of equation (28) for $v = (0,\dots,0)$ is

$$\mathsf{E}(G^{(0,\dots,0)} \mid X^{[\mathcal{J}]}=\ell) = \int g^{(0,\dots,0)} p_\ell(g)\mu_g(dg) = \int p_\ell(g)\mu_g(dg) = 1 \tag{29}$$

Using equation (28), we may obtain for every $j \in \mathcal{J}$ and every $l \in [1..L_j]$:

$$\begin{aligned}
&\sum_k \lambda^k_{jl}\,\mathsf{E}(G^{v+\mathbf{1}_k} \mid X=\ell) = \sum_k \lambda^k_{jl}\int g^{v+\mathbf{1}_k} p_\ell(g)\mu_g(dg) = \\
&\quad \sum_k \int g^v (g_k\lambda^k_{jl}) p_\ell(g)\mu_g(dg) = \int g^v \Big(\sum_k g_k\lambda^k_{jl}\Big)\frac{\prod_{j'\notin\mathcal{J}}\sum_k g_k\lambda^k_{j'\ell_{j'}}}{M_\ell(\mu_\beta)}\mu_g(dg) = \\
&\quad \frac{M_{\ell+\mathbf{l}_j}(\mu_\beta)}{M_\ell(\mu_\beta)}\int g^v \frac{\big(\sum_k g_k\lambda^k_{jl}\big)\prod_{j'\notin\mathcal{J}}\sum_k g_k\lambda^k_{j'\ell_{j'}}}{M_{\ell+\mathbf{l}_j}(\mu_\beta)}\mu_g(dg) = \\
&\quad \frac{M_{\ell+\mathbf{l}_j}(\mu_\beta)}{M_\ell(\mu_\beta)}\mathsf{E}(G^v \mid X=\ell+\mathbf{l}_j)
\end{aligned} \tag{30}$$

By multiplying both sides of (30) by $M_\ell(\mu_\beta)$ one obtains:

$$\sum_k \lambda^k_{jl} \cdot \Big(M_\ell(\mu_\beta) \cdot \mathsf{E}(G^{v+\mathbf{1}_k} \,|\, X=\ell) \Big) = M_{\ell+\boldsymbol{l}_j}(\mu_\beta) \cdot \mathsf{E}(G^v \,|\, X=\ell+\boldsymbol{l}_j) \qquad (31)$$

Equation (31) is the main fact that allows us to establish a numerical procedure to estimate conditional expectations. This equation holds for every $J' \geq 0$, $v \in \mathcal{V}[J']$, $\mathcal{J} \subseteq [1..J]$, $\ell \in \mathcal{L}^{[\mathcal{J}]}$, $j \in \mathcal{J}$, and $l \in [1..L_j]$.

Although equation (31) holds for every J' and $\mathcal{J}$, the most important case is $J' + |\mathcal{J}| < J$: as we shall see in section 7, only conditional moments $\mathsf{E}(G^v \,|\, X=\ell)$, $v \in \mathcal{V}[J']$, $\ell \in \mathcal{L}^{[\mathcal{J}]}$, with $J' + |\mathcal{J}| \leq J$ may be identified from data.

6.3 Conditional Variance

To make use of conditional expectations, one would like to know variance of G conditional on outcomes of measurements. It is not hard to express variance via conditional moments:

$$\mathsf{D}_k(G \,|\, X=\ell) = \int (g_k - \mathsf{E}_k(G \,|\, X=\ell))^2 \, p_\ell(g) \, \mu_g(dg) =$$
$$\mathsf{E}(G^{\mathbf{2}_k} \,|\, X=\ell) - \mathsf{E}^2(G^{\mathbf{1}_k} \,|\, X=\ell) \qquad (32)$$

As we shall show below, $\mathsf{E}(G^{\mathbf{2}_k} \,|\, X=\ell)$ can be identified only for ℓ having at least two components equal 0; thus, the same condition applies to identifiability of $\mathsf{D}(G \,|\, X=\ell)$.

6.4 Change of Basis

Let $\Lambda' = \Lambda A$ be another basis of Q. Here A is nonsingular $K \times K$ matrix,

$$A = \begin{pmatrix} a^1_1 & \dots & a^K_1 \\ & \vdots & \\ a^1_K & \dots & a^K_K \end{pmatrix}, \quad \text{and} \quad A^{-1} = \begin{pmatrix} \bar{a}^1_1 & \dots & \bar{a}^K_1 \\ & \vdots & \\ \bar{a}^1_K & \dots & \bar{a}^K_K \end{pmatrix} \qquad (33)$$

As it was mentioned above, if a vector $\beta \in Q$ has coordinates g in basis Λ, then it has coordinates $g' = A^{-1}g$ in basis Λ'. Thus, A^{-1} is a matrix of transition from coordinates g to coordinates g'. A question of interest is how the moments of G are changed under this transition.

We start with moments M_w for $w \in \mathcal{W}$. Let M'_w be a moment calculated in coordinates g'. Then:

$$M'_w = M'_{(k_1,\dots,k_J)} = \int g'_{k_1} \dots g'_{k_J} \mu'_g(dg') =$$
$$\int (\bar{a}^1_{k_1} g_1 + \dots + \bar{a}^K_{k_1} g_K) \dots (\bar{a}^1_{k_J} g_1 + \dots + \bar{a}^K_{k_J} g_K) \mu_g(dg) =$$
$$\sum_{m_1=1}^K \cdots \sum_{m_J=1}^K \bar{a}^{m_1}_{k_1} \dots \bar{a}^{m_J}_{k_J} \int g_{m_1} \dots g_{m_J} \mu_g(dg) =$$
$$\sum_{m_1=1}^K \cdots \sum_{m_J=1}^K \bar{a}^{m_1}_{k_1} \dots \bar{a}^{m_J}_{k_J} M_{(m_1,\dots,m_J)} \qquad (34)$$

which suggests that $\{M_w\}_{w\in\mathcal{W}}$ is a covariant tensor of rank J. Employing Einstein's convention for summation, (34) may be rewritten,

$$M'_{(k_1,\dots,k_J)} = \bar{a}_{k_1}^{m_1}\dots\bar{a}_{k_J}^{m_J} M_{(m_1,\dots,m_J)} \tag{35}$$

Tensor $\{M_w\}_{w\in\mathcal{W}}$ is symmetric, and $\{M_v\}_{v\in\mathcal{V}}$ is a set of its essential components (as for any $w\in v$, $M_w = M_v$.) Transformation rules for M_v have form:

$$M'_v = \frac{1}{C_v}\sum_{v'}\left(\sum_{w\in v}\sum_{w'\in v'} \bar{a}_{w_1}^{w'_1}\dots\bar{a}_{w_J}^{w'_J}\right) M_{v'} \tag{36}$$

For the general case of conditional moments of arbitrary order, one obtains

$$\mathsf{E}(G'^v \mid X=\ell) = \frac{1}{C_v}\sum_{v'}\left(\sum_{w\in v}\sum_{w'\in v'} \bar{a}_{w_1}^{w'_1}\dots\bar{a}_{w_{J'}}^{w'_{J'}}\right) \mathsf{E}(G^{v'} \mid X=\ell) \tag{37}$$

Here $v\in\mathcal{V}[J']$ for some J', v' ranges over $\mathcal{V}[J']$, w and w' are restricted to the set $\mathcal{W}[J'] = \{(w_1,\dots,w_{J'}) \mid w_j\in[1..K]\}$, and $w\in v$ means "for every k, w contains exactly v_k components equal to k."

7 Main System of Equations

Consider a system of equations,

$$\begin{cases} \sum_k \alpha_{jl}^k h_\ell^{v+1_k} = h_{\ell+1_j}^v, & J'\in[0..J-1],\ v\in\mathcal{V}[J'],\\ & \mathcal{J}\subseteq[1..J] : |\mathcal{J}|>J',\ \ell\in\mathcal{L}^{[\mathcal{J}]},\\ & j\in\mathcal{J},\ l\in[1..L_j] \\ h_\ell^{(0,\dots,0)} = M_\ell, & \ell\in\mathcal{L}^0 \\ \sum_{v\in\mathcal{V}[J']} C_v h_{(0,\dots,0)}^v = 1, & J'\in[0..J] \end{cases} \tag{38}$$

with respect to unknowns α_{jl}^k and h_ℓ^v.

Equations (31) and (29) together with proposition 6.4 give us

Theorem 7.1 *Let $\{M_\ell\}_{\ell\in\mathcal{L}^0}$ be a set of ℓ-moments of distribution μ_β, which satisfies (G1) and (G2). Let also $\{\lambda^k\}_k$, $\lambda^k = (\lambda_{jl}^k)_{jl}$ be some basis of the support of μ_β, and $\mathsf{E}(G^v \mid X=\ell)$ be conditional moments calculated with respect to this basis.*

Then $\alpha_{jl}^k = \lambda_{jl}^k$ and $h_\ell^v = M_\ell\cdot\mathsf{E}(G^v\mid X=\ell)$ give a solution of system (38).

In other words, all values we are interested in are solutions of system (38). Below we establish sufficient conditions for the case when (38) has *only* such solutions.

For the sake of convenience, we (abusing language) shall speak about "solution $\alpha^1,\dots,\alpha^K$," having in mind "there exist h_ℓ^v such that $\alpha^1,\dots,\alpha^K$ together with h_ℓ^v compose a solution."

Let $\alpha^1,\dots,\alpha^K$, $\alpha^k = (\alpha_{jl}^k)_{jl}$, and h_ℓ^v be a solution of (38). Let $\alpha'^1,\dots,\alpha'^K$ be any set of vectors such that $\mathrm{Lin}(\alpha'^1,\dots,\alpha'^K) = \mathrm{Lin}(\alpha^1,\dots,\alpha^K)$. In this case there exists a nonsingular

$K \times K$ matrix $A = (a_k^{k'})_{k'k}$ such that $(\alpha'^1, \ldots, \alpha'^K) = (\alpha^1, \ldots, \alpha^K)A$. Let $A^{-1} = (\bar{a}_k^{k'})_{k'k}$. By straightforward computation one can show that $\alpha'^1, \ldots, \alpha'^K$ together with

$$h_\ell'^{v} = \frac{1}{C_v} \sum_{v'} \left(\sum_{w \in v} \sum_{w' \in v'} \bar{a}_{w_1}^{w'_1} \ldots \bar{a}_{w_{J'}}^{w'_{J'}} \right) h_\ell^{v}$$

is also a solution of (38).

Thus, we can speak about *space of solutions* $\mathrm{Lin}(\alpha^1, \ldots, \alpha^K)$. Note that at this point we have no arguments for uniqueness of the space of solutions; moreover, we cannot even claim that every space of solutions have the same dimension K. In fact, in general, the space of solutions is not unique. However, in the presence of sufficient conditions that we establish below, the space of solutions is unique.

Consider equations from the first group of (38) for $v = (0, \ldots, 0)$ and $\ell = (0, \ldots, 0)$:

$$\left\{ \sum_k \alpha_{jl}^k h_{(0,\ldots,0)}^{\mathbf{1}_k} = h_{\mathbf{l}_j}^{(0,\ldots,0)}, \qquad j \in [1..J],\ l \in [1..L_j] \right. \tag{39}$$

and substitute values for $h_{\mathbf{l}_j}^{(0,\ldots,0)}$ from the second group of (38):

$$\left\{ \sum_k \alpha_{jl}^k h_{(0,\ldots,0)}^{\mathbf{1}_k} = M_{\mathbf{l}_j}, \qquad j \in [1..J],\ l \in [1..L_j] \right. \tag{40}$$

As $h_{(0,\ldots,0)}^{\mathbf{1}_k}$ do not depend on j and l, we obtain

Proposition 7.2 *$(M_{\mathbf{l}_j})_{jl} \in \mathrm{Lin}(\alpha^1, \ldots, \alpha^K)$ for every solution $\alpha^1, \ldots, \alpha^K$ of (38).*

In other words, vector $(M_{\mathbf{l}_j})_{jl}$ belongs to every space of solutions.

Applying similar considerations to the case $\ell = \mathbf{l}'_{j'}$ for some $j' \in [1..J]$, $l' \in [1..L_{j'}]$, we obtain:

$$\left\{ \sum_k \alpha_{jl}^k h_{\mathbf{l}'_{j'}}^{\mathbf{1}_k} = M_{\mathbf{l}'_{j'} + \mathbf{l}_j}, \qquad j \neq j',\ l \in [1..L_j] \right. \tag{41}$$

In system (41) we have equations not for all j, l but only for those in which $j \neq j'$. Thus, (41) does not give us a vector from a solution space. However, it allows us to claim that for every j', l', a vector $(M_{\mathbf{l}'_{j'} + \mathbf{l}_j})_{jl: j \neq j'}$ (having $\sum_{j \neq j'} L_j$ components) may be extended (by adding $L_{j'}$ components) to a $|L|$-dimensional vector that belongs to $\mathrm{Lin}(\alpha^1, \ldots, \alpha^K)$.

In general, for every $\ell \in \mathcal{L}^0 \setminus \mathcal{L}$ we have:

$$\left\{ \sum_k \alpha_{jl}^k h_\ell^{\mathbf{1}_k} = M_{\ell + \mathbf{l}_j}, \qquad \ell_j = 0,\ l \in [1..L_j] \right. \tag{42}$$

and thus we obtain further incomplete vectors that may be completed to vectors belonging to $\mathrm{Lin}(\alpha^1, \ldots, \alpha^K)$.

Let us write vector $(M_{\mathbf{l}_j})_{jl}$ together with incomplete vectors $(M_{\mathbf{l}'_{j'} + \mathbf{l}_j})_{jl: j \neq j'}$, etc., as columns of a matrix, with places for which we do not have moments filled by question marks. We refer to this incomplete matrix as *moment matrix*. The moment matrix contains a column for every $\ell \in \mathcal{L}^0 \setminus \mathcal{L}$. Figure 1 gives an example of (part of) a moment matrix for the case $J = 3$, $L_1 = L_2 = L_3 = 2$. Columns in this matrix correspond to $\ell = (000)$, (100), (200), (010), (020), (001), (002), (110); other columns are not shown.

For the moment matrix M, let its completion $\bar{M}$ be a matrix obtained from M by replacing question marks by arbitrary numbers. The above considerations give us:

$$\begin{pmatrix} M_{(100)} & ? & ? & M_{(110)} & M_{(120)} & M_{(101)} & M_{(102)} & ? & \cdots \\ M_{(200)} & ? & ? & M_{(210)} & M_{(220)} & M_{(201)} & M_{(202)} & ? & \cdots \\ M_{(010)} & M_{(110)} & M_{(210)} & ? & ? & M_{(011)} & M_{(012)} & ? & \cdots \\ M_{(020)} & M_{(120)} & M_{(220)} & ? & ? & M_{(021)} & M_{(022)} & ? & \cdots \\ M_{(001)} & M_{(101)} & M_{(201)} & M_{(011)} & M_{(021)} & ? & ? & M_{(111)} & \cdots \\ M_{(002)} & M_{(102)} & M_{(202)} & M_{(012)} & M_{(022)} & ? & ? & M_{(112)} & \cdots \end{pmatrix}$$

Figure 1: Example of moment matrix

Theorem 7.3 *Let distribution μ_β satisfy (G1) and (G2). Then its moment matrix has a completion $\bar{M}$ such that* $\operatorname{rank}(\bar{M}) \le K$.

One may extend definition of rank to incomplete matrices by setting it equal to the maximal size of nonzero minor, which contains only known moments (i.e. does not contain question marks.) It is easy to see that for every completion $\bar{M}$ of M, inequality $\operatorname{rank}(M) \le \operatorname{rank}(\bar{M})$ holds. Thus,

Corollary 7.4 *Let distribution μ_β satisfy (G1) and (G2). Then* $\operatorname{rank}(M) \le K$.

For $\mathcal{K} \subseteq \mathcal{L}^0 \setminus \mathcal{L}$, let $M[\mathcal{K}]$ denote a matrix consisting of those columns of moment matrix M that correspond to elements of $\mathcal{K}$.

Now we are ready to formulate the third assumption regarding distribution μ_β:

(G3) There exists a subset of column indices $\mathcal{K} \subseteq \mathcal{L}^0 \setminus \mathcal{L}$ such that:

(a) For every two completions of moment matrix $\bar{M}'$ and $\bar{M}''$ satisfying $\operatorname{rank}(\bar{M}') \le K$ and $\operatorname{rank}(\bar{M}'') \le K$, the equality $\bar{M}'[\mathcal{K}] = \bar{M}''[\mathcal{K}]$ holds.

(b) Let $\bar{M}$ be any completion of moment matrix satisfying $\operatorname{rank}(\bar{M}) \le K$. Then $\operatorname{rank}(\bar{M}[\mathcal{K}]) = K$.

Note that when (G3) holds, $\bar{M}[\mathcal{K}]$ is uniquely defined.

Theorem 7.5 *Let distribution μ_β satisfy (G1), (G2), and (G3). Then for every solution of system (38)* $\operatorname{Lin}(\alpha^1, \ldots, \alpha^K) = \operatorname{Lin}(\bar{M}[\mathcal{K}])$ *(where* $\operatorname{Lin}(\bar{M}[\mathcal{K}])$ *is a linear subspace of* $\mathbb{R}^{|L|}$ *spanned by columns of* $\bar{M}[\mathcal{K}]$.*)*

Proof. By theorem 7.3, for every solution of (38) there exists a completion $\bar{M}'$ of M such that $\operatorname{Lin}(\bar{M}') \subseteq \operatorname{Lin}(\alpha^1, \ldots, \alpha^K)$. Then $\operatorname{rank}(\bar{M}') = \dim(\operatorname{Lin}(\bar{M}')) \le \dim(\operatorname{Lin}(\alpha^1, \ldots, \alpha^K)) \le K$. Thus, by (G3), $\bar{M}'[\mathcal{K}] = \bar{M}[\mathcal{K}]$, and consequently $\operatorname{Lin}(\bar{M}[\mathcal{K}]) \subseteq \operatorname{Lin}(\bar{M}')$. Further, as $\dim(\operatorname{Lin}(\bar{M}[\mathcal{K}])) = \operatorname{rank}(\bar{M}[\mathcal{K}]) = K$, we obtain $\operatorname{Lin}(\alpha^1, \ldots, \alpha^K) = \operatorname{Lin}(\bar{M})$. ■

Corollary 7.6 *Let distribution μ_β satisfy (G1), (G2), and (G3). Then:*

(a) *To obtain a solution of (38), it is enough to take* $\alpha^1, \ldots, \alpha^K$ *equal to any basis of* $\bar{M}[\mathcal{K}]$ *(e.g., equal to any K linearly independent columns of* $\bar{M}[\mathcal{K}]$*.)*

(b) *Any other solution* $\alpha'^1, \ldots, \alpha'^K$ *is obtained from the above one by multiplying it by nonsingular* $K \times K$ *matrix.*

(c) *Every solution* $\alpha^1, \ldots, \alpha^K$ *is a basis of* Q*, a support of* μ_β.

Proof. (a) and (b) are obvious.

To prove (c), consider that by theorem 7.1, every basis of Q is a solution of (38). By (b), all solutions are bases of the same linear subspace of $\mathbb{R}^{|L|}$. Thus, every solution is a basis for Q. ■

By theorem 7.5 and its corollary, assumption (G3) is sufficient to identify a support of μ_β. It looks like it is close to a necessary condition, as in many cases where (G3) is violated we were able to construct a different distribution μ'_β, which has the same ℓ-moments as μ_β (and therefore μ'_β is indistinguishable from μ_β based on available observations.) However, the exact formulation of necessary conditions for identifiability of support of μ_β is an open question.

To verify whether condition (G3) holds, it is enough to analyze the moment matrix. Numerous practical methods might be suggested to do such verification. Without going into details, we demonstrate by example one possibility.

Example 7.7 Consider a case $J = 3$, $L_1 = L_2 = L_3 = 2$; thus, $\mathbb{R}^{|L|} = \mathbb{R}^6$. Consider a distribution μ_β concentrated in three points, $\beta^{(1)}$, $\beta^{(2)}$, and $\beta^{(3)}$, with every point having the probability $\frac{1}{3}$ (see figure 2). As $\beta^{(3)} = \frac{1}{2}\beta^{(1)} + \frac{1}{2}\beta^{(2)}$ and $\{\beta^{(1)}, \beta^{(2)}, \beta^{(3)}\} \in \mathrm{Supp}(\mu_\beta)$, (G2) is satisfied for $K = 2$.

The moment matrix M of this distribution (which corresponds to moment matrix in figure 1) is shown on figure 2.

A submatrix of M consisting of rows 3 and 4 and columns 1 and 2 is nonsingular, and therefore x and y such that

$$\text{column1} \cdot x + \text{column2} \cdot y = \text{column7}$$

are uniquely defined; they are $x = \frac{131}{160}$ and $y = -\frac{99}{160}$. This allows construction of the only possible completion of column 7, which is shown in figure 2.

Thus, column 1 and (completed) column 7 give a basis for the support of μ_β. It is easy to see that $\mathrm{Lin}(\text{column1}, \text{column7}) = \mathrm{Lin}(\beta^{(1)}, \beta^{(2)}, \beta^{(3)})$, as one would expect.

Vectors "column1" and "completed column7" do not satisfy condition (Λ_0). To obtain a basis satisfying (Λ_0), one can take $\alpha^1 = \text{column1}$ and $\alpha^2 = \text{column7} \cdot \frac{40}{19}$. Vectors α^1 and α^2 are shown on figure 2.

(Calculations for this and subsequent examples were done with Waterloo Maple ™ v.7.00.) ■

The second question is whether h^ν_ℓ may be uniquely determined from (38) given a solution $\alpha^1, \ldots, \alpha^K$. In general, the answer is negative: not all h^ν_ℓ may be uniquely determined. However, a number of the most important values always may be determined uniquely, as the following theorem shows.

$$\beta^{(1)} = \begin{pmatrix} \beta^{(1)}_{11} \\ \beta^{(1)}_{12} \\ \beta^{(1)}_{21} \\ \beta^{(1)}_{22} \\ \beta^{(1)}_{31} \\ \beta^{(1)}_{32} \end{pmatrix} = \begin{pmatrix} 1 \\ 0 \\ \frac{1}{3} \\ \frac{2}{3} \\ \frac{4}{5} \\ \frac{1}{5} \end{pmatrix}, \quad \beta^{(2)} = \begin{pmatrix} \beta^{(2)}_{11} \\ \beta^{(2)}_{12} \\ \beta^{(2)}_{21} \\ \beta^{(2)}_{22} \\ \beta^{(2)}_{31} \\ \beta^{(2)}_{32} \end{pmatrix} = \begin{pmatrix} \frac{1}{9} \\ \frac{8}{9} \\ \frac{3}{5} \\ \frac{2}{5} \\ \frac{1}{4} \\ \frac{3}{4} \end{pmatrix}, \quad \beta^{(3)} = \begin{pmatrix} \beta^{(3)}_{11} \\ \beta^{(3)}_{12} \\ \beta^{(3)}_{21} \\ \beta^{(3)}_{22} \\ \beta^{(3)}_{31} \\ \beta^{(3)}_{32} \end{pmatrix} = \begin{pmatrix} \frac{5}{9} \\ \frac{4}{9} \\ \frac{7}{15} \\ \frac{8}{15} \\ \frac{21}{40} \\ \frac{19}{40} \end{pmatrix}$$

$$M = \begin{pmatrix} \frac{5}{9} & ? & ? & \frac{89}{405} & \frac{136}{405} & \frac{403}{1080} & \frac{197}{1080} & ? & \cdots \\ \frac{4}{9} & ? & ? & \frac{20}{81} & \frac{16}{81} & \frac{41}{270} & \frac{79}{270} & ? & \cdots \\ \frac{7}{15} & \frac{89}{405} & \frac{20}{81} & ? & ? & \frac{397}{1800} & \frac{443}{1800} & ? & \cdots \\ \frac{8}{15} & \frac{136}{405} & \frac{16}{81} & ? & ? & \frac{137}{450} & \frac{103}{450} & ? & \cdots \\ \frac{21}{40} & \frac{403}{1080} & \frac{41}{270} & \frac{397}{1800} & \frac{137}{450} & ? & ?[\leftarrow \frac{191}{960}] & \frac{151}{1080} & \cdots \\ \frac{19}{40} & \frac{197}{1080} & \frac{79}{270} & \frac{443}{1800} & \frac{103}{450} & ? & ?[\leftarrow \frac{53}{192}] & \frac{259}{3240} & \cdots \end{pmatrix}$$

$$\alpha^1 = \begin{pmatrix} \alpha^1_{11} \\ \alpha^1_{12} \\ \alpha^1_{21} \\ \alpha^1_{22} \\ \alpha^1_{31} \\ \alpha^1_{32} \end{pmatrix} = \begin{pmatrix} \frac{5}{9} \\ \frac{4}{9} \\ \frac{7}{15} \\ \frac{8}{15} \\ \frac{21}{40} \\ \frac{19}{40} \end{pmatrix}, \quad \alpha^2 = \begin{pmatrix} \alpha^2_{11} \\ \alpha^2_{12} \\ \alpha^2_{21} \\ \alpha^2_{22} \\ \alpha^2_{31} \\ \alpha^2_{32} \end{pmatrix} = \begin{pmatrix} \frac{197}{513} \\ \frac{316}{513} \\ \frac{443}{855} \\ \frac{412}{855} \\ \frac{191}{456} \\ \frac{265}{456} \end{pmatrix}$$

Figure 2: Illustration to example 7.7

Theorem 7.8 *Let $\alpha^1, \ldots, \alpha^K$ be a solution of (38), and let set of index pairs $j_1 l_1, \ldots, j_K l_K$, with $l_k \in [1..L_{j_k}]$, be chosen so that the matrix $(\alpha^{k'}_{j_k l_k})_{k'k}$ is nonsingular (this is always possible as* $\operatorname{rank}(\alpha^1, \ldots, \alpha^K) = K$.*) Let $\mathfrak{J}_0 = \{j_1, \ldots, j_K\}$ (note that $|\mathfrak{J}_0|$ may be less than K.) Then:*

(a) For every $\mathfrak{J}$ such that $\mathfrak{J}_0 \subseteq \mathfrak{J}$, for every $\ell \in \mathfrak{L}^{[\mathfrak{J}]}$, and for every $k \in [1..K]$, the conditional expectation $\mathsf{E}(G_k \mid X = \ell)$ is uniquely defined.

(b) Let, in addition, there exist $j_0 \notin \mathfrak{J}_0$ and $l_0 \in [1..L_{j_0}]$ such that every $K \times K$ submatrix of $(K+1) \times K$ matrix $(\alpha^{k'}_{j_k l_k})_{k' \in [1..K], k \in [0..K]}$ is nonsingular. Then for every $\mathfrak{J}$ such that $\mathfrak{J}_0 \cup \{j_0\} \subseteq \mathfrak{J}$, for every $\ell \in \mathfrak{L}^{[\mathfrak{J}]}$, and for every $k \in [1..K]$, the conditional variance $\mathsf{D}(G_k \mid X = \ell)$ is uniquely defined.

Proof. (a) Consider a subsystem of (38):

$$\left\{ \sum_{k'} \alpha^{k'}_{j_k l_k} h^{\mathbf{1}_{k'}}_{\ell} = M_{\ell+(l_k)_{(j_k)}}, \qquad k = 1,\ldots,K \right.$$

By theorem 7.1, $h^{\mathbf{1}_{k'}}_{\ell} = M_\ell \cdot \mathsf{E}(G_k \mid X = \ell)$ is a solution of this system, and by assumption of the theorem, there are no other solutions.

(b) By part (a) of the theorem, for every $k_0 \in [1..K]$ and every $k \in [1..K]$, values $h^{\mathbf{1}_{k_0}}_{\ell+(l_k)_{(j_k)}}$ are uniquely determined from (38). Now consider a subsystem of (38):

$$\left\{ \sum_{k'} \alpha^{k'}_{j_k l_k} h^{\mathbf{1}_{k_0}+\mathbf{1}_{k'}}_{\ell} = h^{\mathbf{1}_{k_0}}_{\ell+(l_k)_{(j_k)}}, \qquad k = 1,\ldots,K;\ k_0 = 1,\ldots,K \right.$$

By theorem 7.1, $h^{\mathbf{1}_{k_0}+\mathbf{1}_{k'}}_{\ell} = M_\ell \cdot \mathsf{E}(G^{\mathbf{1}_{k_0}+\mathbf{1}_{k'}} \mid X = \ell)$ is a solution of this system, and by assumption of the theorem, there are no other solutions. This is enough to calculate $\mathsf{D}(G_k \mid X = \ell)$ using formula (32). ■

Example 7.9 We continue example 7.7. Consider a subsystem of (38):

$$\begin{cases} \alpha^1_{21} h^{(1,0)}_{(1,0,0)} + \alpha^2_{21} h^{(0,1)}_{(1,0,0)} = M_{(1,1,0)} \\ \alpha^1_{22} h^{(1,0)}_{(1,0,0)} + \alpha^2_{22} h^{(0,1)}_{(1,0,0)} = M_{(1,2,0)} \end{cases}, \quad \text{or} \quad \begin{cases} \frac{7}{15} h^{(1,0)}_{(1,0,0)} + \frac{443}{855} h^{(0,1)}_{(1,0,0)} = \frac{89}{405} \\ \frac{8}{15} h^{(1,0)}_{(1,0,0)} + \frac{412}{855} h^{(0,1)}_{(1,0,0)} = \frac{136}{405} \end{cases}$$

Solving this system gives

$$h^{(1,0)}_{(1,0,0)} = \frac{131}{99}, \qquad h^{(0,1)}_{(1,0,0)} = -\frac{76}{99}$$

and, as $h^{\nu}_{\ell} = M_\ell \cdot \mathsf{E}(G^{\nu} \mid X = \ell)$,

$$\mathsf{E}(G^{(1,0)} \mid X = (1,0,0)) = \frac{131}{55}, \qquad \mathsf{E}(G^{(0,1)} \mid X = (1,0,0)) = -\frac{76}{55}$$

Considering similar subsystems, one obtains, in particular,

$$h^{(1,0)}_{(1,0,1)} = \frac{3089}{2970}, \quad h^{(0,1)}_{(1,0,1)} = -\frac{2641}{3960}, \quad h^{(1,0)}_{(1,0,2)} = \frac{841}{2970}, \quad h^{(0,1)}_{(1,0,2)} = -\frac{133}{3960}$$

Substituting these values into subsystems,

$$\begin{cases} \alpha^1_{31} h^{(2,0)}_{(1,0,0)} + \alpha^2_{31} h^{(1,1)}_{(1,0,0)} = h^{(1,0)}_{(1,0,1)} \\ \alpha^1_{32} h^{(2,0)}_{(1,0,0)} + \alpha^2_{32} h^{(1,1)}_{(1,0,0)} = h^{(1,0)}_{(1,0,2)} \end{cases}, \quad \begin{cases} \alpha^1_{31} h^{(1,1)}_{(1,0,0)} + \alpha^2_{31} h^{(0,2)}_{(1,0,0)} = h^{(0,1)}_{(1,0,1)} \\ \alpha^1_{32} h^{(1,1)}_{(1,0,0)} + \alpha^2_{32} h^{(0,2)}_{(1,0,0)} = h^{(0,1)}_{(1,0,2)} \end{cases}$$

one finds,

$$h^{(2,0)}_{(1,0,0)} = \frac{3323}{726}, \qquad h^{(1,1)}_{(1,0,0)} = -\frac{7087}{2178}, \qquad h^{(0,2)}_{(1,0,0)} = \frac{1895}{726}$$

and thus,

$$\mathsf{E}(G^{(2,0)} \mid X = (1,0,0)) = \frac{9969}{1210}, \qquad \mathsf{E}(G^{(0,2)} \mid X = (1,0,0)) = \frac{1083}{242}$$

This allows us calculate conditional variances (using formula (32)):

$$\mathsf{D}(G_1 \mid X = (1,0,0)) = \frac{15523}{6050}, \qquad \mathsf{D}(G_2 \mid X = (1,0,0)) = \frac{15523}{6050}$$

Table 1 summarize conditional expectations and conditional variances that may be calculated in our example. Although all values are *exact* rational numbers, we used decimal notation to make comparison of values easier. We also put standard deviations in the table instead of variances.

As we have mentioned, there are many choices for the basis of the support of distribution μ_β. Another possibility is to take $\{\beta^{(1)}, \beta^{(2)}\}$ as a basis. The result of calculations in this basis is given in table 2. One can see that, although numbers are different, their relative position remains the same. ■

Table 1: Conditional expectations and standard deviations calculated in basis $\{\alpha^1, \alpha^2\}$.

ℓ	$\mathsf{E}(G_1 \mid X = \ell)$	$\sigma(G_1 \mid X = \ell)$	$\mathsf{E}(G_2 \mid X = \ell)$	$\sigma(G_2 \mid X = \ell)$
$(1,0,0)$	2.3818	1.6018	-1.3818	1.6018
$(2,0,0)$	-0.7273	1.2214	1.7273	1.2214
$(0,1,0)$	0.5065	2.0571	0.4935	2.0571
$(0,2,0)$	1.4318	2.0709	-0.4318	2.0709
$(0,0,1)$	1.9048	1.9122	-0.9048	1.9122
$(0,0,2)$	0.0000	1.8642	1.0000	1.8642

Table 2: Conditional expectations and standard deviations calculated in basis $\{\beta^{(1)}, \beta^{(2)}\}$.

ℓ	$\mathsf{E}(G_1 \mid X = \ell)$	$\sigma(G_1 \mid X = \ell)$	$\mathsf{E}(G_2 \mid X = \ell)$	$\sigma(G_2 \mid X = \ell)$
$(1,0,0)$	0.7667	0.3091	0.2333	0.3091
$(2,0,0)$	0.1667	0.2357	0.8333	0.2357
$(0,1,0)$	0.4048	0.3970	0.5952	0.3970
$(0,2,0)$	0.5833	0.3997	0.4167	0.3997
$(0,0,1)$	0.6746	0.3690	0.3254	0.3690
$(0,0,2)$	0.3070	0.3598	0.6930	0.3598

Remark 7.10 The standard deviations in the above example are relatively large. This is direct consequence of the fact that in this example we have too small number of measurements. When number of measurements increases, the standard deviation becomes smaller and smaller. ■

Remark 7.11 Theorem 7.8 guarantees that it is always possible to find $J-K$ measurements such that expectations of G conditional on outcomes of these measurements may be uniquely determined from the system (38). The possibility of determining conditional variances is not guaranteed by this theorem, however. In many practical cases that we have investigated, conditions of the part (b) of theorem 7.8 are satisfied, and conditional variances can be found (as in example 7.9.) The exact conditions for determinability of conditional variances is an open question. ■

Remark 7.12 By computations similar to used in (17) and (21), one obtains for every family of $J' \leq J$ index pairs $j_1 l_1, \ldots, j_{J'} l_{J'}$ with $l_p \in [1..L_{j_p}]$ (j_p is not necessarily different from $j_{p'}$ for $p \neq p'$)

$$\int \beta_{j_1 l_1} \cdot \ldots \cdot \beta_{j_{J'} l_{J'}} \mu_\beta(d\beta) = \sum_{v \in \mathcal{V}[J']} M_v(\mu_g) \cdot \tilde{\Lambda}(v, j_1, l_1, \ldots, j_{J'}, l_{J'})$$

where $\tilde{\Lambda}(v, j_1, l_1, \ldots, j_{J'}, l_{J'})$ depends only on λ^k_{jl}. Thus, if the system (38) allows unique determination of all unknowns h^v_ℓ, all moments of order up to J of μ_β can be identified. This is the case, for instance, in the example 7.9. In general case, only a subset of these moments may be identified (that corresponds to the subset of identifiable conditional expectations according to the theorem 7.8). ■

The theorems of this section describe the set of identifiable invariants of GoM models. These invariants are values that may be uniquely derived from the system (38). According to the theorem 7.5, the supporting plane of the mixing measure μ_β is an invariant under conditions (G1)–(G3). According to the theorem 7.8, many conditional expectations are invariants as well.

The existence of these invariants is very important for applications of the model. Namely, the supporting plane may be considered as a phase space, and conditional expectations give positions of individuals in this phase space. In other words, the GoM models allow us to derive low-dimensional continuous hidden states from high-dimensional categorical observations.

8 Numerical Procedure

We have established a number of *precise* relations between values of interest (i.e. expectation and variance of hidden random vector G conditional on outcomes of measurements) and moments of (unknown) distribution μ_β, which are directly estimable from observations. The most important of these relations are given by equations (31), and by system of equations (38). These relations suggest a numerical procedure for estimation of values of interest.

As was mentioned above, sample frequencies f_ℓ are consistent estimators for moments $M_\ell(\mu_\beta)$. Thus, applying the least squares method to the system

$$\begin{cases} \sum_k \alpha^k_{jl} h^{\nu+1_k}_{\ell} = h^{\nu}_{\ell+1_j}, & J' \in [0..J-1],\ \nu \in \mathcal{V}[J'], \\ & \mathfrak{J} \subseteq [1..J] : |\mathfrak{J}| > J',\ \ell \in \mathcal{L}^{[\mathfrak{J}]}, \\ & j \in \mathfrak{J},\ l \in [1..L_j] \\ h^{(0,\dots,0)}_{\ell} = f_{\ell}, & \ell \in \mathcal{L}^0 \\ \sum_{\nu \in \mathcal{V}[J']} C_\nu h^{\nu}_{(0,\dots,0)} = 1, & J' \in [0..J] \end{cases} \tag{43}$$

one obtains consistent estimators for a basis $\{\lambda^k\}_k$ and conditional expectations of G.

The consistency of estimators obtained from (43) is almost straightforward corollary to consistency of estimators f_ℓ. The rate of convergence is a more delicate question (as a rate of convergence of f_ℓ depends on ℓ,) and deserves separate investigation.

Theorem 7.5 suggests another, two-step way for finding solutions of (43). On the first step, one finds a basis from frequency matrix (i.e. moment matrix with frequencies substituted for moments.) After a basis is obtained, (43) turns out to be a linear system with respect to h^{ν}_{ℓ}. This way requires significantly less computations, but its convergence properties have to be more carefully investigated.

One question regarding the numerical procedure is the choice of value of K for which system (43) should be solved. Theorem 7.3 and its corollary suggest that one has to take K equal to the rank of the frequency matrix (modulo possible deviations of frequencies from the true moments.)

Another question is how a numerical algorithm has to deal with is nonuniqueness of basis $\{\lambda^k\}_k$. In general, there are K^2 degrees of freedom in choice of a basis. Imposing condition (Λ_0) reduces this number to $K(K-1)$. One can consider additional restrictions on choice of basis:

(Λ_1) For every k, unconditional expectation $\mathsf{E}_k(G)$ equals $\frac{1}{K}$.

(Λ_2) The map H_Λ is isometry of $\bar{P}_g$ and $\bar{P}_\beta$ (with respect to euclidean distance.)

The first one corresponds to restricting transformations of $\bar{P}_g$, described by matrix A (introduced in section 4) to those having the "center" point $(\frac{1}{K},\dots,\frac{1}{K})$ of $\bar{P}_g$ fixed. The second restriction guarantees that variances do not depend on the choice of basis, and variances calculated in g-space coincide with variances calculated in β-space.

The nonuniqueness of the basis should be considered as an advantage of the method, as it gives to the applied researcher the ability to choose a "point of view" on the problem that better expose its most interesting aspects — like a choice of appropriate reference frame in mechanics allows significantly simplify the problem.

Estimation of variances is another source of problems. Formula (32) is of theoretical importance, as it demonstrates that we have enough information to estimate variances. However, it hardly can be used for numerical computations as it involves *differences* of values that we can only approximately estimate. We are working on finding a better way to estimate variances.

9 Conclusion

We developed a novel approach to analysis of categorical data based on considering distribution laws of observed random variables as realizations of another random variable, i.e. considering the observed distribution law as a mixture of independent distribution laws. This starting point leads to a fruitful development.

In the present article, we were able to obtain system of equations (38) and establish its properties in theorems 7.1–7.8. This provides a base for an efficient numerical procedure.

We also believe that the approach in general, and our results regarding system (38) in particular, may be successfully applied in other domains of statistics, especially in latent structure analysis.

References

[1] Barndorff-Nielsen, O. (1965) Identifiability of mixtures of exponential families. *Journal of Mathematical Analysis and Applications* , **12**, 115-121.

[2] Bartholomew, D.J., & Knott, M. (1999) *Latent Variable Models and Factor Analysis.* 2nd ed., London: Arnold; New York: Oxford University Press.

[3] Bishop, Y. M. M., Fienberg, S. E., and Holland, P. W. (1975) *Discrete Multivariate Analysis.* MIT Press, Cambridge, MA.

[4] Clogg, C.C. (1995) *Latent Class Models.* In "Handbook of Statistical Modeling for the Social and Behavioral Sciences", Arminger, G., Clogg, C.C., & Sobel, M.E., eds., New York: Plenum Press, 311–360.

[5] Erosheva, E.A. (2002) Grade of Membership and Latent Structure Models With application to Disability Survey Data. Ph.D. Thesis, Department of Statistics, Carnegie Mellon University. Available at http://www.stat.cmu.edu/ fienberg/NLTCS_Models/Erosheva-thesis-2002.pdf

[6] Erosheva, E. A. (2003) *Bayesian Estimation of the Grade of Membership Model.* Volume 7 of "Bayesian Statistics," Oxford University Press, Oxford, UK.

[7] Erosheva, E. A. and Fienberg, S. E. (2004) Bayesian mixed membership models for soft classification. Working Paper No. 40, Center for Statistics and the Social Sciences, University of Washington.

[8] Goodman, L. A. (1978) Analyzing Qualitative/Categorical Data: Log-linear Models and Latent-Structure Analysis. Abt Books, Cambridge, MA.

[9] Haberman, S.J. (1995) Book review of 'Statistical Applications Using Fuzzy Sets', by Kenneth G. Manton, Max A. Woodbury, and H. Dennis Tolley. *Journal of the American Statistical Association*, **90** (431), 1131–1133.

[10] Heinen, T. (1996) Latent class and discrete latent trait models: similarities and differences. Thousand Oaks, Calif.: Sage Publications.

[11] Langeheine, R. and Rost, J., editors (1988) *Latent Trait and Latent Class Models.* Plenum Press, New York.

[12] Lazarsfeld, P. F. (1950a) The Interpretation and Computation of some Latent Structures. Volume IV of "Studies in Social Psychology in World War II," chapter 11, pages 413-472, Princeton University Press, Princeton, NJ.

[13] Lazarsfeld, P. F. (1950b) *The Logical and Mathematical Foundations of Latent Structure Analysis.* Volume IV of "Studies in Social Psychology in World War II," chapter 10, pages 362-412. Princeton University Press, Princeton, NJ.

[14] Lazarsfeld, P. F. and Henry, N. W. (1968) *Latent Structure Analysis.* Houghton Mifflin Company, Boston, MA.

[15] Lindsay, B. G. (1983) The geometry of mixture likelihood: A general theory. *The Annals of Statistics*, **11** (1), 86-94.

[16] Lindsay, B. G. (1995) *Mixture models: Theory, Geometry and Applications.* Volume 5 of "NSF-CBMS Regional Conference Series in Probability and Statistics," Institute of Mathematical Statistics, Hayward, CA.

[17] Manton, K. G., Woodbury, M. A., and Tolley, H. D. (1994) *Statistical applications using fuzzy sets.* John Wiley and Sons, New York.

[18] Neyman, J., & Scott, E.L. (1948) Consistent Estimates Based on Partially Consistent Observations. *Econometrica*, **16** (1), 1-32.

[19] Newcomb, S. (1886) A generalized theory of combination of observations so as to obtain the best result. *American Journal of Mathematics*, **8**, 343-366.

[20] Patil, G. P. and Bildikar, S. (1966) Identifiability of countable mixtures of discrete probability distributions using methods of infinite matrices. *Proceeding of Cambridge Philosophical Society*, **62**, 485-494.

[21] Pearson, K. (1894) Contribution to the mathematical theory of evolution. *Philosophical Transactions of the Royal Society A* , **185**, 71-110.

[22] Tallis, G. M. (1969) The identifiability of mixtures of distributions. *Journal of Applied Probability*, **6**, 389-398.

[23] Tallis, G. M. and Chesson, P. (1982) Identifiability of mixtures. *Journal of the Australian Mathematical Society (Series A)* , **32**(3), 339-348.

[24] Teicher, H. (1960) On the mixture of distributions. *The Annals of Mathematical Statistics*, **31** (1), 55-73.

[25] Teicher, H. (1961) Identifiability of mixtures. *The Annals of Mathematical Statistics* , **32** (1), 244-248.

[26] Teicher, H. (1963) Identifiability of finite mixtures. *The Annals of Mathematical Statistics*, **34** (4), 1265-1269.

[27] Woodbury, M., & Clive, J. (1974) Clinical pure types as a fuzzy partition. *Journal of Cybernetics* **4**, 111–121.

[28] Woodbury, M. A., Clive, J., and Garson, A. (1978) Mathematical typology: a grade of membership technique for obtaining disease definition. *Computers and Biomedical Research*, **11**, 277-298.

[29] Yakowitz, S. J. and Spragins, J. D. (1968) In the identifiability of finite mixture. *The Annals of Mathematical Statistics* , **39** (1), 209-214.

In: Focus on Probability Theory
Editor: Louis R. Velle, pp. 27-41
ISBN 1-59454-474-3

Chapter 2

SUMMABILITY AND ABSOLUTE SUMMABILITY FUNCTIONS FOR SUMMABILITY TRANSFORMS OF RANDOM VARIABLES

Rohitha Goonatilake*
Department of Mathematical and Physical Sciences
Texas A&M International University
Laredo, TX 78041-1900

Abstract

In this paper, we discuss the strong regularity concept of Lorentz [8] applied to the convolution method of summability. Later, we obtain the summability functions and absolute summability functions of this method.

1 Introduction

The methods which sum all almost convergent sequences are called strongly regular. We will show in section II that the matrix transformation corresponding to the regular convolution method generated by an independent and identically distributed sequence of aperiodic nonnegative integer-valued random variables with finite third moment and positive variance is strongly regular.

Summability functions [8] in some sense determine the strength of the regularity of method for bounded sequences. It may also be used to show that Tauberian conditions of a certain kind may not be improved. Under the existence of first three moments, in section III, it is shown that $\Omega(n) = o(\sqrt{n})$ are the summability functions for the convolution methods, thus extending some of previously known results for other methods such as Borel and Euler. The optimality of class of summability functions is also ascertained to show that all functions of the form, $\Omega(n) = o(\sqrt{n})$ are only these functions are summability functions for the methods $C(p,q)$ with some moment conditions. We conclude this paper with a discussion of absolute summability functions for this method.

*E-mail address: harag@tamiu.edu

The discussion will now revolve around the following types of convolution summability methods [12] and [14]. Let $Y, X_1, X_2, \cdots$ be a sequence of independent non-negative integer valued random variables such that Y has probability function q and the $X_i's$ are identically distributed with probability function p. Let $S_0 = Y$ and $S_n = Y + X_1 + \cdots + X_n$ for $n \geq 1$. Let $\{p_j\}_{j\geq 0}$ and $\{q_j\}_{j\geq 0}$ be the weights of probability distributions of $X_1, X_2, \cdots$ and Y respectively. The nth row k^{th} column entry of the convolution summability matrix C is the probability $C_{n,k} = P(S_n = k)$. The method C is regular if and only if $P(X_1 = 0) < 1$ [7]. Some classical summability methods are examples of the method C. For instance, when $Y = 0$ and $X_1 \sim \text{Binomial}(1,1)$, then C becomes the Euler method denoted by E_r. When $Y \sim X_1 \sim \text{Poisson}(1)$ we get the Borel matrix method. When $Y \sim \text{Geometric}(1-r)$ and $X_1 \sim Y + 1$, then we get the Taylor method. And when $Y \sim X_1 \sim \text{Geometric}(1-r)$ we get the Meyer-König method. We shall call C a convolution method and when $Y = 0$ with probability 1, it is called the random-walk method. The method C can be extended to non-identically distributed random variables (for example, Jakimovski family of summability methods [14]); however, it will serve our purpose adequately for the time being, as it is. The regular convolution summability matrix $\{C_{n,k\geq 0}\}$, referred to everywhere in this paper has the above construction with appropriate moment conditions and in section IV with finite moment generating function of $\{X_i, i \geq 1\}$.

2 Strong Regularity

Given below is a definition of almost convergence of a sequence, which is as we see, a generalization of ordinary convergence. Even more interesting to note that would be a sequence which is not convergent but which is almost convergent.

Definition 1 A bounded sequence $\{x_i\}_{i\geq 0}$ is called almost convergent, if there is a number s such that

$$\lim_{p\to\infty} \frac{x_n + x_{n+1} + \cdots + x_{n+p-1}}{p} = s \text{ holds uniformly in } n.$$

We denote s by $\text{Lim}\, x_n$.

Example 1 For a complex z on the boundary of the unit circle $\text{Lim}\, z^n = 0$ holds everywhere except for $z = +1$, as follows from

$$\frac{1}{p}(z^n + z^{n+1} + \cdots + z^{n+p-1}) = \frac{z^n}{p}\left(\frac{1 - z^p}{1 - z}\right).$$

Example 2 For a positive integer n, the sequence $x_n = n^{-2}$ is almost convergent for $\alpha > 1$.

We now use the following theorem of Lorentz [8].

Theorem 1 *(Lorentz [8]) In order that regular matrix method (transformation) $A = \{a_{n,k}\}$ sum all almost convergent sequences, it is necessary and sufficient that*

$$\lim_{n\to\infty} \sum_{k=0}^{\infty} |a_{n,k} - a_{n,k+1}| = 0.$$

In view of the above result, we now give below the definition for strong regularity of an almost convergent sequence.

Definition 2 A summability method A is called strongly regular, if for any almost convergent sequence $\{x_i\}_{i\geq 0}$ with $\text{Lim}\, x_n = l$, we have $\lim_{n\to\infty}(Ax)_n = l$.

In [8], Lorentz showed that the Cesàro method C_α of order $\alpha > 0$ and the Euler method E_r, with parameter r, are strongly regular. In an attempt to generalize these results, we will prove that the random-walk method is strongly regular for a probability function with finite third moment. Then using this result, we show that the convolution summability method is also strongly regular.

Theorem 2 *Let $\xi_1, \xi_2, \xi_3, \cdots$ be an i. i. d. sequence of aperiodic nonnegative integer-valued random variables with finite third moment and positive variance. Then the matrix transformation corresponding to the random-walk method of the above sequence of random variables is strongly regular.*

Prior to the proof of this theorem, we need an important theorem due to Bikelis and Jasjunas [1], which gives the rate of convergence for the central limit theorem:

Theorem 3 *(Bikelis & Jasjunas [1]) For a sequence $\{\xi_i\}_{i\geq 1}$, i. i. d. aperiodic nonnegative integer-valued random variables with mean μ, positive variance σ^2, and finite third moment, the following holds:*

$$\sum_{j=-\infty}^{\infty}(1+|\frac{j-n\mu}{\sigma\sqrt{n}}|^3)|P(S_n=j)-\frac{1}{\sigma(2\pi n)^{\frac{1}{2}}}\exp\{-\frac{1}{2}(j-n\mu)^2/(n\sigma^2)\}|=O(n^{-1/2}),$$

where $S_n = \xi_1 + \cdots + \xi_n$.

Proof of theorem 2

As suggested in Theorem 1, we now consider

$$\sum_{k=0}^{\infty}|a_{n.k+1}-a_{n.k}|=\sum_{k=0}^{\infty}|P(S_n=k+1)-P(S_n=k)|,\text{ where } S_n=\sum_{i=0}^{n}\xi_i \text{ and } \xi_0=0.$$

If the mean of ξ_i is μ and standard deviation of ξ_i is σ, we write

$$\begin{aligned}
&\sum_{k=0}^{\infty}|P(S_n=k+1)-P(S_n=k)|\\
=&\sum_{k=0}^{\infty}|\{P(S_n=k+1)-\frac{1}{\sigma(2\pi n)^{\frac{1}{2}}}\exp\{-\frac{1}{2}(k+1-n\mu)^2/(n\sigma^2)\}\}\\
&+\{\frac{1}{\sigma(2\pi n)^{\frac{1}{2}}}\exp\{-\frac{1}{2}(k-n\mu)^2/(n\sigma^2)\}-P(S_n=k)\}\\
&+\{\frac{1}{\sigma(2\pi n)^{\frac{1}{2}}}\exp\{-\frac{1}{2}(k+1-n\mu)^2/(n\sigma^2)\}-\frac{1}{\sigma(2\pi n)^{\frac{1}{2}}}\exp\{-\frac{1}{2}(k-n\mu)^2/(n\sigma^2)\}\}|
\end{aligned}$$

$$\leq \sum_{k=0}^{\infty} |P(S_n = k+1) - \frac{1}{\sigma(2\pi n)^{\frac{1}{2}}} \exp\{-\frac{1}{2}(k+1-n\mu)^2/(n\sigma^2)\}|$$
$$+\sum_{k=0}^{\infty} |\frac{1}{\sigma(2\pi n)^{\frac{1}{2}}} \exp\{-\frac{1}{2}(k-n\mu)^2/(n\sigma^2)\} - P(S_n = k)|$$
$$+\sum_{k=0}^{\infty} |\frac{1}{\sigma(2\pi n)^{\frac{1}{2}}} \exp\{-\frac{1}{2}(k+1-n\mu)^2/(n\sigma^2)\} - \frac{1}{\sigma(2\pi n)^{\frac{1}{2}}} \exp\{-\frac{1}{2}(k-n\mu)^2/(n\sigma^2)\}|.$$

Subject to the finiteness of the third moment, considering the fact that $(1+|\frac{j-n\mu}{\sqrt{n\sigma^2}}|^3) \geq 1$ and restricting the values of j, for $j \geq 0$, we obtain from Bikelis and Jasjunas theorem 3 that

$$\sum_{j=0}^{\infty} |P(S_n = j) - \frac{1}{\sigma(2\pi n)^{\frac{1}{2}}} \exp\{-\frac{1}{2}(j-n\mu)^2/(n\sigma^2)\}| = O(n^{-1/2}).$$

This implies that the first two sums of the above are in fact of $O(n^{-1/2})$.

For the last sum, we look at the telescopic series in the following form, noting that as $k \longrightarrow \infty$, the terms increase until $k = n\mu$ and then decreases to 0 thereafter:

$$\frac{1}{\sigma(2\pi n)^{\frac{1}{2}}}\{(\exp\{-\frac{1}{2}(1-n\mu)^2/(n\sigma^2)\} - \exp\{-\frac{1}{2}(0-n\mu)^2/(n\sigma^2)\})$$
$$+(\exp\{-\frac{1}{2}(2-n\mu)^2/(n\sigma^2)\} - \exp\{-\frac{1}{2}(1-n\mu)^2/(n\sigma^2)\}) + \cdots$$
$$+\cdots+(\ldots\ldots\ldots\ldots - \ldots\ldots\ldots\ldots) + \cdots$$
$$+\cdots+(\exp\{-\frac{1}{2}(i+1-n\mu)^2/(n\sigma^2)\} - \exp\{-\frac{1}{2}(i-n\mu)^2/(n\sigma^2)\}) + \cdots$$
$$+\cdots+(\ldots\ldots\ldots\ldots - \ldots\ldots\ldots\ldots) + \cdots$$
$$+\cdots+(\exp\{-\frac{1}{2}(n\mu-n\mu)^2/(n\sigma^2)\} - \exp\{-\frac{1}{2}(n\mu-1-n\mu)^2/(n\sigma^2)\})\}$$
$$+\frac{1}{\sigma(2\pi n)^{\frac{1}{2}}}\{(\exp\{-\frac{1}{2}(n\mu+1-n\mu)^2/(n\sigma^2)\} - \exp\{-\frac{1}{2}(n\mu+2-n\mu)^2/(n\sigma^2)\})$$
$$+(\exp\{-\frac{1}{2}(n\mu+2-n\mu)^2/(n\sigma^2)\} - \exp\{-\frac{1}{2}(n\mu+3-n\mu)^2/(n\sigma^2)\}) + \cdots$$
$$+\cdots+(\ldots\ldots\ldots\ldots - \ldots\ldots\ldots\ldots) + \cdots$$
$$+\cdots+(\exp\{-\frac{1}{2}(j-n\mu)^2/(n\sigma^2)\} - \exp\{-\frac{1}{2}(j+1-n\mu)^2/(n\sigma^2)\}) + \cdots$$
$$+\cdots+(\ldots\ldots\ldots\ldots - \ldots\ldots\ldots\ldots)\};$$

for $0 \leq i \leq n\mu - 1$ and $n\mu + 1 < j < \infty$.

Since $\lim_{j\to\infty} \exp\{-\frac{1}{2}(j-n\mu)^2/(n\sigma^2)\} = 0$, the above series sums to

$$\frac{1}{\sigma(2\pi n)^{\frac{1}{2}}}\{-e^{-\frac{1}{2}n^2\mu^2/n\sigma^2} + 1 + e^{-\frac{1}{2}/n\sigma^2}\}$$

$$= \frac{1}{\sigma(2\pi n)^{\frac{1}{2}}}\{-e^{-n\mu^2/2\sigma^2} + 1 + e^{-1/2n\sigma^2}\} = O(1/\sqrt{n}).$$

This together with above leads to the fact that

$$\sum_{k=0}^{\infty} |P(S_n = k+1) - P(S_n = k)| = O(1/\sqrt{n}).$$

Now, by the Lorentz criteria, we have the strong regularity of the random-walk method. ▮

We will use the above result to prove the following generalization.

Theorem 4 *Let $Y, \{X_i, i \geq 1\}$ be independent and let $\{X_i, i \geq 1\}$ be identically distributed aperiodic nonnegative integer-valued random variables with finite third moment. Let C be the resulting convolution summability method. Then the following are equivalent.*

(i) $Var(X_1) > 0$,

(ii) C is strongly regular.

Proof.

We will first show that (i) implies (ii).

Let $\{q_j\}$ and $\{p_j\}$ be the probability weights associated with random variables Y and $\{X_i, i \geq 1\}$. The weight of the convolution summability method is

$$C_{n,k} = P(Y + S_n = k) = \sum_{j=0}^{k} q_j\{P(S_n = k-j)\}, \text{ where } S_n = \sum_{i=1}^{n} X_i.$$

Now,

$$\begin{aligned}
&\sum_{k=0}^{\infty} |C_{n,k+1} - C_{nk}| \\
= &\sum_{k=0}^{\infty} |\sum_{j=0}^{k+1} q_j P(S_n = k+1-j) - \sum_{j=0}^{k} q_j P(S_n = k-j)| \\
= &\sum_{k=0}^{\infty} |\sum_{j=0}^{k} q_j P(S_n = k+1-j) - \sum_{j=0}^{k} q_j P(S_n = k-j) + q_{k+1} P(S_n = 0)| \\
\leq &\sum_{k=0}^{\infty} \sum_{j=0}^{k} q_j |P(S_n = k+1-j) - P(S_n = k-j)| + \sum_{k=0}^{\infty} q_{k+1} P(S_n = 0) \\
\leq &\sum_{j=0}^{\infty} q_j \sum_{k=j}^{\infty} |P(S_n = k+1-j) - P(S_n = k-j)| + P(S_n = 0)
\end{aligned}$$

as $\sum_{k=0}^{\infty} q_k = 1$.

Change of summation index $k - j \to k$ gives

$$\begin{aligned}
&\sum_{k=0}^{\infty} |C_{n,k+1} - C_{nk}| \\
\leq &\sum_{j=0}^{\infty} q_j \sum_{k=0}^{\infty} |P(S_n = k+1) - P(S_n = k)| + (p_0)^n \\
\leq &\sum_{k=0}^{\infty} |P(S_n = k+1) - P(S_n = k)| + (p_0)^n
\end{aligned}$$

as $\sum_{k=0}^{\infty} q_k = 1$. As already seen in the previous proof, the first term is of $O(n^{-1/2})$, provided that the $\{X_i\}'s$ have finite third moment and positive variance, whereas the second term tends to 0. Since we assumed that $\text{Var}(X_1) > 0$, it must be that $p_0 < 1$. Hence

$$\sum_{k=0}^{\infty} |C_{n,k+1} - C_{n,k}| \to^{n} 0 \qquad (n \to \infty).$$

To prove that (ii) implies (i), assume that (ii) holds and (i) fails. When $\text{Var}(X_1) = 0$ there exists a nonnegative integer m such that $P(X_1 = m) = 1$. Hence,

$$C_{n,j} = P(Y + S_n = j) = P(Y = j - nm)$$
$$= \begin{cases} 0 & \text{if} \quad j < nm \\ q_{j-nm} & \text{if} \quad j \geq nm. \end{cases}$$

Therefore,

$$\sum_{j=0}^{\infty} |C_{n,j+1} - C_{n,j}| \geq \sum_{j=0}^{\infty} |q_{j+1} - q_j| \neq 0,$$

as $\sum_{j=0}^{\infty} q_j = 1$, and $q_i \geq 0$ for $i \geq 0$. This contradiction gives the result. ▌

Remark 1 It should be noted that with the condition $p_0 < 1$, Khan [7] proved the regularity of the convolution summability method. Our condition $\text{Var}(X_1) > 0$ implies that $p_0 < 1$. Furthermore, it follows as a result of Theorem 4 that Taylor and Meyer-König methods are strongly regular.

3 Summability Functions

The concept of summability functions was introduced by Lorentz [8]. There are many uses of the summability functions. Summability functions, in some sense, determine the strength of the regularity of the method for bounded sequences and also may be used to show that Tauberian conditions of a certain kind cannot be improved.

Definition 3 The class $\mathcal{U}$ is the set of regular matrix methods $A = \{a_{n,k}\}$ for which

$$\lim_{n \to \infty} \{\max_k |a_{n,k}|\} = 0$$

is fulfilled.

Every regular convolution summability method satisfies this property, so they form a subset of $\mathcal{U}$. For,

$$C_{n,k} = \sum_{j=0}^{k} q_j \{P(\sum_{i=1}^{n} X_i = k - j)\},$$

where $\{C_{n,k}\}$ are the convolution summability weights under consideration. Now with $\mu = EX_1$ and $0 < \sigma^2 = \text{Var}(X_1) < \infty$,

$$C_{n,k} = \sum_{j=0}^{k} q_j [(\frac{1}{2\pi n\sigma^2})^{1/2} \exp\{-\frac{1}{2}(k - j - n\mu)^2/n\sigma^2\} + o(1/\sqrt{n})],$$

uniformly in $k-j$. Since $\sum_{j=0}^{\infty} q_j = 1$ and $q_j \geq 0$, $\forall\, j \geq 0$, we have

$$\max_k |C_{n,k}| = O(\frac{1}{\sqrt{n}}) \to^n 0.$$

The methods of the class $\mathcal{U}$ are characterized by the fact that they all possess summability functions. We now give the precise definition of the summability function.

Definition 4 Given a matrix $A = \{a_{n,k}\}$, a nonnegative sequence $\Omega(n)$ that increases to ∞ is called a summability function for A if $s_n \to 0$ (A) holds whenever $s_n = O(1)$ and $A(n,s) = \sum_{\nu \leq n,\, s_\nu \neq 0} 1 \leq \Omega(n)$. The sequence $A(n,s)$ is sometimes called a counting function of the sequence $\{s_n\}$.

Theorem 5 *(Lorentz [8]) The condition* $\lim_{m\to\infty}\{\max_n |a_{m,n}|\} = 0$ *is necessary as well as sufficient for the existence of an integer-valued function* $\Omega(n)$ *that increases to* ∞*, such that every bounded sequence* $x = \{x_n\}$ *for which the indices* n_ν*, with* $x_{n_\nu} \neq 0$ *have a counting function* $A(n,x) \leq \Omega(n)$ *is* $A-$*summable to zero.*

The following theorem gives sufficient conditions under which the existence of summability functions can be determined.

Theorem 6 *(Lorentz [8]) Let* $A = \{a_{n,k}\}$ *be a regular matrix summability method. If the integer-valued function of* k, $f(k)$, *is such that* $0 < f(k) \uparrow \infty$ *and if*

$$\sum_{k=0}^{\infty} f(k)|a_{n,k} - a_{n,k+1}| = O(1),$$

then every nonnegative sequence $\Omega(n)$ *that increases to* ∞, $\Omega(n) = o(f(n))$ *is a summability function for* A.

We make use of the above theorem to show that $\Omega(k) = o(\sqrt{k})$ is a summability function for the convolution summability methods generated by a sequence of aperiodic nonnegative integer-valued random variables with finite third moment. The optimality of the summability function so obtained is also ascertained. In this connection, we shall show that all functions $\Omega(n) = o(\sqrt{n})$ and only these functions are summability functions for the methods $\{C_{n,k}\}_{n,k\geq 1}$. In [8], it has been shown that all functions of the form $\Omega(n) = o(\sqrt{n})$ and only these functions are summability functions for the Euler method E_α with $\alpha > 0$. We use this fact in the proof of the following theorem.

There are cases where Theorem 6 does not give all summability functions. For an example, the Nölund method N_p, with $p_n = 1/(n+1)$; Theorem 6 gives that $\Omega(n) = o(\log(n))$ are its summability functions. However, it is known that any function $\Omega(n) = O(n^{\varepsilon_n})$ with $0 < \varepsilon_n \to 0$ is a summability function for N_p (p. 61, [13]).

The following theorem provides summability functions for a convolution summability method.

Theorem 7 *Let* $Y, \{X_i, i \geq 1\}$ *be independent with* $E(Y) = \mu_Y < \infty$*, and* $\{X_i, i \geq 1\}$ *be identically distributed aperiodic nonnegative integer-valued random variables with finite third* *moment and positive variance. Then for the matrix transformation corresponding to the regular convolution summability method $C = \{C_{n,k}\}$, any function $0 < \Omega(n) \uparrow \infty$ of the form $\Omega(n) = o(\sqrt{n})$ gives a summability function of C. Furthermore, $0 < \Omega(n) = o(\sqrt{n})$ with $\Omega(n) \uparrow \infty$ are the only functions which are summability functions over the class of regular convolution methods under consideration.*

Proof.

The weight of the convolution summability method is given by

$$C_{n,k} = P(Y + S_n = k) = \sum_{j=0}^{k} q_j \{P(S_n = k - j)\} \text{ where } S_n = \sum_{i=1}^{n} X_i.$$

Now consider,

$$\begin{aligned}
&\sum_{k=0}^{\infty} \sqrt{k} |C_{n,k+1} - C_{nk}| \\
&= \sum_{k=0}^{\infty} \sqrt{k} |\sum_{j=0}^{k+1} q_j P(S_n = k + 1 - j) - \sum_{j=0}^{k} q_j P(S_n = k - j)| \\
&= \sum_{k=0}^{\infty} \sqrt{k} |\sum_{j=0}^{k} q_j P(S_n = k + 1 - j) - \sum_{j=0}^{k} q_j P(S_n = k - j) + q_{k+1} P(S_n = 0)| \\
&\le \sum_{k=0}^{\infty} \sqrt{k} \sum_{j=0}^{k} q_j |P(S_n = k + 1 - j) - P(S_n = k - j)| + \sum_{k=0}^{\infty} \sqrt{k} q_{k+1} P(S_n = 0) \\
&\le \sum_{j=0}^{\infty} q_j \sum_{k=j}^{\infty} \sqrt{k} |P(S_n = k + 1 - j) - P(S_n = k - j)| + P(S_n = 0) \sum_{k=0}^{\infty} \sqrt{k} q_{k+1}.
\end{aligned}$$

Making the change of summation index $k - j \to k$ in the first sum and $k + 1 \to k$ in the second sum, we obtain

$$= \sum_{j=0}^{\infty} q_j \sum_{k=0}^{\infty} \sqrt{k + j} |P(S_n = k + 1) - P(S_n = k)| + P(S_n = 0) \sum_{k=1}^{\infty} \sqrt{k - 1} q_k.$$

Since the sequence $\{q\}$ has a finite first moment, say μ_Y, it follows that

$$\sum_{k=0}^{\infty} \sqrt{k} |C_{n,k+1} - C_{nk}| \le \sum_{j=0}^{\infty} q_j \sum_{k=0}^{\infty} (\sqrt{k} + \sqrt{j}) |P(S_n = k + 1) - P(S_n = k)| + \mu_Y (p_0)^n.$$

Note that the last term tends to 0, since $\text{Var}(X_1) > 0$, which implies that $p_0 < 1$.

$$\sum_{k=0}^{\infty} \sqrt{k} |C_{n,k+1} - C_{nk}|$$

$$\le \mu_Y \sum_{k=0}^{\infty} |P(S_n = k + 1) - P(S_n = k)| + \sum_{k=0}^{\infty} \sqrt{k} |P(S_n = k + 1) - P(S_n = k)| + o(1).$$

The first sum also tends to 0, since $\mathrm{Var}(X_1) > 0$ is a necessary and sufficient condition for the strong regularity of the convolution summability method. Now what remains is to show that

$$\sum_{k=0}^{\infty} \sqrt{k}|P(S_n = k+1) - P(S_n = k)| = O(1).$$

We begin with the following.

$$\begin{aligned}\sqrt{k} &= \sqrt{(k-n\mu)+n\mu} \leq \sqrt{|k-n\mu|+n\mu} \leq \sqrt{|k-n\mu|}+(n\mu)^{1/2} \\ &\leq |\frac{k-n\mu}{\sigma\sqrt{n}}|^{1/2}(\sigma\sqrt{n})^{1/2}+(n\mu)^{1/2}.\end{aligned}$$

With the assumption of the finiteness of the third moment of the sequence $\{p_j\}$ of i. i. d. random variables Theorem 3 of Bikelis and Jasjunas [1] gives

$$\sum_{j=-\infty}^{\infty}(1+|\frac{j-n\mu}{\sigma\sqrt{n}}|^3)|P(S_n = j) - \frac{1}{\sigma(2\pi n)^{\frac{1}{2}}}\exp\{-\frac{1}{2}(j-n\mu)^2/(n\sigma^2)\}| = O(n^{-1/2}),$$

where $S_n = X_1 + X_2 + \cdots + X_n$, mean of $\{X_i\}$ is μ and standard deviation of $\{X_i\}$ is σ. Then, we have

$$\begin{aligned}&\sum_{k=0}^{\infty}\sqrt{k}|P(S_n = k+1) - P(S_n = k)| \\ \leq\ &(\sigma\sqrt{n})^{1/2}\sum_{k=0}^{\infty}|\frac{k-n\mu}{\sigma\sqrt{n}}|^{1/2}|P(S_n = k+1) - P(S_n = k)| \\ &+(n\mu)^{1/2}\sum_{k=0}^{\infty}|P(S_n = k+1) - P(S_n = k)| = \sum_1 + \sum_2 \text{ say.}\end{aligned}$$

We have already shown in the proof of Theorem 2 that $\sum_2 = O(1)$.

For $\sum_1$, we will proceed as follows:

$$\begin{aligned}\sum_1 \leq\ &(\sigma\sqrt{n})^{1/2}\sum_{k=0}^{\infty}|\frac{k-n\mu}{\sigma\sqrt{n}}|^{1/2}|P(S_n = k+1) \\ &-\frac{1}{\sigma(2\pi n)^{\frac{1}{2}}}\exp\{-\frac{1}{2}(k+1-n\mu)^2/(n\sigma^2)\}| \\ &+(\sigma\sqrt{n})^{1/2}\sum_{k=0}^{\infty}|\frac{k-n\mu}{\sigma\sqrt{n}}|^{1/2}|\frac{1}{\sigma(2\pi n)^{\frac{1}{2}}}\exp\{-\frac{1}{2}(k-n\mu)^2/(n\sigma^2)\} - P(S_n = k)| \\ &+(\sigma\sqrt{n})^{1/2}\sum_{k=0}^{\infty}|\frac{k-n\mu}{\sigma\sqrt{n}}|^{1/2}\frac{1}{\sigma(2\pi n)^{\frac{1}{2}}}|\exp\{-\frac{1}{2}(k+1-n\mu)^2/(n\sigma^2)\} \\ &-\exp\{-\frac{1}{2}(k-n\mu)^2/(n\sigma^2)\}|.\end{aligned}$$

Note that

$$\left(1+|\frac{k-n\mu}{\sigma\sqrt{n}}|^3\right) > |\frac{k-n\mu}{\sigma\sqrt{n}}|^{1/2} \text{ for all } k \geq 0 \text{ and all } n \geq 0.$$

This shows that the first two sums are of order $O(1)$ as we expected. Now we consider the last sum:

$$(\sigma\sqrt{n})^{1/2}\frac{1}{\sigma(2\pi n)^{\frac{1}{2}}}\sum_{k=0}^{\infty}|\frac{k-n\mu}{\sigma\sqrt{n}}|^{1/2}|\exp\{-\frac{1}{2}(k+1-n\mu)^2/(n\sigma^2)\}$$
$$-\exp\{-\frac{1}{2}(k-n\mu)^2/(n\sigma^2)\}|.$$

Let $t_{n,k}=\frac{k-n\mu}{\sigma\sqrt{n}}$ and let $\Delta t_{n,k}=t_{n,k+1}-t_{n,k}=\frac{1}{\sigma\sqrt{n}}$. The last sum is

$$\begin{aligned}
&\frac{\sqrt{\sigma\sqrt{n}}}{\sqrt{2\pi}}\sum_{k=0}^{\infty}\frac{1}{\sigma\sqrt{n}}|t_{n,k}|^{1/2}\left|e^{-\frac{1}{2}t_{n,k+1}^2}-e^{-\frac{1}{2}t_{n,k}^2}\right|\\
=\;&\frac{\sqrt{\sigma\sqrt{n}}}{\sqrt{2\pi}}\sum_{k=0}^{\infty}\left(t_{n,k+1}-t_{n,k}\right)|t_{n,k}|^{1/2}\left|e^{-\frac{1}{2}\left(t_{n,k}+\frac{1}{\sigma\sqrt{n}}\right)^2}-e^{-\frac{1}{2}t_{n,k}^2}\right|\\
=\;&\frac{\sqrt{\sigma\sqrt{n}}}{\sqrt{2\pi}}\sum_{k=0}^{\infty}\left(\Delta t_{n,k}\right)|t_{n,k}|^{1/2}\left|e^{-\frac{1}{2}t_{n,k}^2}\left(e^{-\frac{1}{2}\left(2t_{n,k}\Delta t_{n,k}+\left(\Delta t_{n,k}\right)^2\right)}-1\right)\right|\\
=\;&\sqrt{\sigma\sqrt{n}}\sum_{k=0}^{\infty}\left(\Delta t_{n,k}\right)|t_{n,k}|^{1/2}\phi\left(t_{n,k}\right)\left|e^{t_{n,k}\Delta t_{n,k}}e^{-\frac{1}{2}\left(\Delta t_{n,k}\right)^2}-1\right|\\
=\;&\sqrt{\sigma\sqrt{n}}\sum_{k=0}^{\infty}\left(\Delta t_{n,k}\right)|t_{n,k}|^{1/2}\phi\left(t_{n,k}\right)\left|\sum_{j=0}^{\infty}\frac{\left(-t_{n,k}\Delta t_{n,k}-\frac{\left(\Delta t_{n,k}\right)^2}{2}\right)^j}{j!}-1\right|\\
\leq\;&\sqrt{\sigma\sqrt{n}}\sum_{k=0}^{\infty}\left(\Delta t_{n,k}\right)|t_{n,k}|^{1/2}\phi\left(t_{n,k}\right)\sum_{j=0}^{\infty}\left(\Delta t_{n,k}\right)\frac{\left(|t_{n,k}|+1\right)^j}{j!}\\
\leq\;&\frac{\sqrt{\sigma\sqrt{n}}}{\sigma\sqrt{n}}\sum_{k=0}^{\infty}\left(\Delta t_{n,k}\right)|t_{n,k}|^{1/2}\phi\left(t_{n,k}\right)e^{1+|t_{n,k}|}\\
\sim\;&\frac{1}{\sqrt{\sigma\sqrt{n}}}\int_{-\infty}^{\infty}|t|^{1/2}e^{|t|+1}\phi(t)dt\\
=\;&O\left(\frac{1}{n^{1/4}}\right).
\end{aligned}$$

This now concludes the proof of the first half of the theorem.

Since E_α with $\alpha>0$ are members of the convolution method and as proved in [8], $\Omega(n)=o(\sqrt{n})$ are the only summability functions of the method E_α, one cannot enlarge the class of summability functions over the space of convolution methods under consideration. This concludes the sharpness of the result. ∎

Remark 2 The summability functions for the following methods were obtained by Lorentz [8], of which $\Omega(n)=o(\sqrt{n})$ for the Euler E_p method agrees with the above theorem.

1. For the $(C,1)$ method; $\Omega(n)=o(n)$. As the methods (C,α) $(\alpha>0)$ and the Abel method $\mathcal{A}$ are equivalent to the $(C,1)$ method for bounded sequences, they also have the same summability functions.

2. For the Euler E_p method; $\Omega(n) = o(\sqrt{n})$.

Let $R_{n,j}$ be the weight of the random-walk method. As usual, by writing μ, σ^2 for the mean and variance of the sequence of i. i. d. random variables with finite third moment, we obtain

$$\max_j |R_{n,j}| = O\Big(\frac{1}{n^{1/2}}\Big) \text{ as } n \longrightarrow \infty$$

as follows from

$$R_{n,j} = P(X_1 + X_2 + \cdots + X_n = j) = \frac{1}{\sigma(2\pi n)^{1/2}} \exp\{\frac{-1}{2}(j - n\mu)^2/n\sigma^2\} + o(1/\sqrt{n})$$

uniformly in j, where mean of $\{X_i\}$ is μ and standard deviation of $\{X_i\}$ is σ [2]. Hence, the set of all random-walk methods with finite third moment is contained in the class $\mathcal{U}$ of the matrix method. The following corollary can be easily drawn from the above theorem.

Corollary 1 *Let $\{X_i, i \geq 1\}$ be independent and identically distributed aperiodic nonnegative integer-valued random variables with finite third moment and positive variance. Then for the matrix transformation corresponding to the regular random-walk method, $\{R_{n,j}\}$; any function $0 < \Omega(n) \uparrow \infty$ of the form $\Omega(n) = o(\sqrt{n})$ gives the possible summability functions. Furthermore, $0 < \Omega(n) = o(\sqrt{n})$ with $\Omega(n) \uparrow \infty$ are the only functions which are summability functions over the class of regular random-walk methods.*

4 Absolute Summability Functions

Definition of the summability functions has been improved by introducing the concept of absolute summability functions.

Definition 5 Let $\Omega(n)$ be a non-decreasing positive function which tends to $+\infty$ with n. We say that $\Omega(n)$ is an absolute summability function of a summability matrix $A = \{a_{n,k}\}_{n,k\geq 0}$, if any bounded sequence $\{f(k),\ k \geq 0\}$ for which $f(k) = 0$ except for a subsequence $\{n_\nu\}$ with the counting function $A(n, f) \leq \Omega(n)$ is absolutely A-summable, that is, $\sum_{n=0}^{\infty} |\sigma_n - \sigma_{n-1}| < +\infty$ for any such sequence, where $\sigma_n = \sum_{k=0}^{\infty} a_{n,k} f(k)$ for $n \geq 0$.

Theorem 7 of Lorentz [10] addresses question of the existence of absolute summability functions.

Theorem 8 *(Lorentz [10]) The method of summation A generated by the matrix $A = \{a_{n,k}\}_{n,k\geq 0}$ for which $\sum_{k=0}^{\infty} |a_{0,k}| < +\infty$ has absolute summability functions if and only if the variation of the $k-$th column $V_k = var_n\ a_{n,k}$ defined by $\sum_{n=0}^{\infty} |a_{n+1,k} - a_{n,k}|$ converges to 0 for $k \longrightarrow \infty$.*

As we will show below, a regular convolution summability method that has been considered in the preceding sections has this structure. Hence, according to theorem 7 of [10], a regular convolution summability method under consideration has absolute summability functions. Since the moment generating function (mgf) may exist for some real arguments but not all, we simply insist that the characteristic function is to be entire (analytic) in the results to follow [11]. For the probabilistic relevance of the mgf condition, see [3].

Theorem 9 *Let $Y, \{X_i, i \geq 1\}$ be independent with $E(Y) = \mu_Y < \infty$, and $\{X_i, i \geq 1\}$ be identically distributed aperiodic nonnegative integer-valued random variables with characteristic function is analytic. The matrix transformation corresponding to the regular convolution summability method $C = \{C_{n,k}\}_{n,k \geq 0}$ has absolute summability functions.*

Proof.
First, we verify that

$$\sum_{k=0}^{\infty} |C_{0,k}| = \sum_{k=0}^{\infty} P(Y = k) = \sum_{k=0}^{\infty} q_k = 1 < \infty.$$

This means that the method already satisfies the condition of hypotheses.
Let $S_n = X_1 + X_2 + \cdots + X_n$ and $S_0 = 0$. We now consider,

$$\begin{aligned}
& \sum_{n=0}^{\infty} |C_{n+1,k} - C_{n,k}| \\
= & \sum_{n=0}^{\infty} |P(Y + S_{n+1} = k) - P(Y + S_n = k)| \\
= & \sum_{n=0}^{\infty} \Big| \sum_{j=0}^{k+1} q_j P(S_n = k+1-j) - \sum_{j=0}^{k} q_j P(S_n = k-j) \Big| \\
\leq & \sum_{n=0}^{\infty} \Big| \sum_{j=0}^{k} |P(S_n = k+1-j) - P(S_n = k-j)| + \sum_{n=0}^{\infty} q_{k+1} P(S_n = 0) \Big) \\
\leq & \sum_{j=0}^{k} q_j \sum_{n=0}^{\infty} |P(S_n = k+1-j) - P(S_n = k-j)| + q_{k+1} \sum_{n=0}^{\infty} (p_0)^n \\
= & \sum_{j=0}^{k} q_j \sum_{n=0}^{\infty} |P(S_n = k+1-j) - P(S_n = k-j)| + q_{k+1} \left(\frac{1}{1-p_0} \right).
\end{aligned}$$

The last term on the right is $o_k(1)$ as the method is regular ($p_0 < 1$,) and $\sum_{k=0}^{\infty} q_k = 1$. The convergence of first (other) sum is evident from Theorem 4 of [6]. Now, we show that this sum is in fact $o_k(1)$, where k denote that order notation has taken as $k \to \infty$. For,

$$I_k = \sum_{j=0}^{k} q_j \sum_{n=0}^{\infty} |P(S_n = k+1-j) - P(S_n = k-j)| + o_k(1),$$

using the Chung-Erdõs inequality cited in page 706 of [6], of the form: If for some integer $k \geq 0, a$, $P(S_k = a)P(S_{k+m} = a+j) > 0$ holds, then for every $\varepsilon > 0$ there exists a $\delta > 0$ such that sufficiently large n,

$$\begin{aligned}
P(S_n = i_n) & \leq (1+\varepsilon)P(S_{n+m} = i_n + j) + e^{-\delta n} \text{ and} \\
P(S_{n+m} = i_n + j) & \leq (1+\varepsilon)P(S_n = i_n) + e^{-\delta n}.
\end{aligned}$$

Thus, for $\delta_1 \neq \delta_2$, we have I_k

$$\leq \sum_{j=0}^{k} q_j \sum_{n=0}^{\infty} |(1+\varepsilon_1)P(S_{n+k-j}=0)+e^{-\delta_1(n+k-j)}-(1+\varepsilon_2)P(S_{n+k-j}=0)$$
$$-e^{-\delta_2(n+k-j)}|+o_k(1)$$
$$\leq \sum_{j=0}^{k} q_j\Big\{|\varepsilon_1-\varepsilon_2|\sum_{n=0}^{\infty}(p_0)^{n+k-j}+\Big|\sum_{n=0}^{\infty}(e^{-\delta_1(n+k-j)}-\sum_{n=0}^{\infty}(e^{-\delta_2(n+k-j)}\Big|\Big\}+o_k(1)$$
$$\leq |\varepsilon_1-\varepsilon_2|(p_0)^k\sum_{j=0}^{k} q_j p_0^{-j}\Big(\frac{1}{1-p_0}\Big)+\sum_{j=0}^{k} q_j e^{\max(\delta_1,\delta_2)j}\Big|\sum_{n=0}^{\infty}\Big\{e^{-\delta_1 n}-e^{-\delta_2 n}\Big\}\Big|$$
$$\Big(e^{-\min(\delta_1,\delta_2)}\Big)^k+o_k(1)$$
$$\leq |\varepsilon_1-\varepsilon_2|(p_0)^k\sum_{j=0}^{k} q_j p_0^{-j}\Big(\frac{1}{1-p_0}\Big)+\sum_{j=0}^{k} q_j e^{\max(\delta_1,\delta_2)j}\Big|\frac{1}{1-e^{-\delta_1}}-\frac{1}{1-e^{-\delta_2}}\Big|$$
$$\Big(e^{-\min(\delta_1,\delta_2)}\Big)^k+o_k(1).$$

This gives $I_k = o_k(1)$ with characteristic function of $\{X_i, i \geq 1\}$ is analytic. ∎

In analogy to Theorem 9 of Lorentz [10], we prove the following generalization for the convolution summability method; which now includes Taylor and Meyer-König summability methods and many others. We will use most of the preliminary facts from [10] and avoid further discussions as we proceed.

Theorem 10 *A function $\Omega(n)$ is an absolute summability function of the regular convolution summability method generated by a sequence of $Y, \{X_i, i \geq 1\}$ of independent with $E(Y) = \mu_Y < \infty$, and $\{X_i, i \geq 1\}$ of identically distributed aperiodic non negative integer-valued random variables with characteristic function is analytic, if and only if*

$$\sum_{n=1}^{\infty} n^{-3/2}\Omega(n) < +\infty.$$

Proof.

The sufficiency of the theorem is proved as follows: Let $\{C_{n,k}\}_{n,k\geq 1}$ be the convolution summability matrix corresponding to the given sequence of random variables. From Theorem 9 the variation of the $k-$th column is

$$V_k = \mathrm{var}_n\, C_{n,k} = \sum_{n=0}^{\infty} |C_{n+1,k} - C_{n,k}| = \alpha(p_0)^k + O\Big(e^{-\beta k}\Big) + \gamma q_{k+1}$$

for some positive constants α, β, and γ as follows from the proof of Theorem 9. If $\{n_\nu\}$ is a sequence of integers with the counting function $\omega(n) \leq \Omega(n)$, we have $\sum_{\nu=1}^{\infty} k_\nu^{-\beta} < \infty$ by Lemma 2 (p. 247 [10]), and noting that $e^{-\beta k_\nu} < k_\nu{}^{-\beta}$, we see that

$$\sum_{\nu=1}^{\infty} \mathrm{var}_n C_{n,k_\nu} = \sum_{\nu=1}^{\infty}\Big\{\alpha(p_0)^{k_\nu} + O\Big(e^{-\beta k_\nu}\Big) + \gamma q_{k_\nu+1}\Big\} < \infty.$$

The later is the necessary and sufficient condition for $\Omega(n)$ to be an absolute summability function of the given matrix method (Theorem 6 of [10]).

For the necessity, using the fact that Euler method E_t, $0 < t < 1$ and the Borel method B are members of the convolution summability methods, it suffices to proceed in the following manner. Suppose the series $\sum_{n=1}^{\infty} n^{-\frac{3}{2}}\Omega(n)$ be divergent. Then taking the Euler method E_t, $0 < t < 1$ or the Borel method B, we have that the series is convergent for either of these methods. This contradiction concludes the assertion. ∎

Most of the discussions and remarks appeared in [8], [9], [10], and [13] now follow without further proofs and hold for the random-walk method and all members of the convolution summability method.

Acknowledgements

This paper stems from the author's Ph. D. dissertation [5] at the Kent State University, Kent, Ohio under Professor M. Kazim Khan. The previous preliminary version of this paper has appeared in [4]. The author wishes to thank him for proposing this topic for further exploration with some pertinent insights to the problem in hand. Various other readers have helped to improve this paper to a greater extent. They are appreciated.

References

[1] Bikelis, A. and Jasjunas, G. : *Limits in the metric of the spaces* L_1, l_1, (in Russian) 1967, 195-218, No. 7, Litovsk. Mat. Sb..

[2] Bingham, N. H. and Maejima, Makoto : *Summability methods and almost sure convergence,* 1985, No. 68, pp. 383-392, Z. Wahrch. Verw. Gebiete.

[3] Csörgő, M. and Révész, P. : *Strong approximations in probability and statistics,* 1981, Academic Press, NY.

[4] Goonatilake, Rohitha : *Some results on summability of random variables,* 2004, Vol. 7, No. 04.3.4, pp. 1-16, J. of Int. Sequen.

[5] Goonatilake, Rohitha : *On probabilistic aspects of summability theory,* Dec. 1997, Department of Mathematics and Computer Sciences, Kent State University, Kent, Ohio 44242.

[6] Kesten, Harry : *Sums of independent random variables-without moments conditions,* 1972, Vol.43, No.3, pp. 701-732, Ann. of Math. Statist..

[7] Khan, M. K. : *Statistical methods in analysis I: Some Tauberian theorems for absolute summability,* Pak. J. Statist., Vol. 7(1)A, 1991, pp. 21-32.

[8] Lorentz, G. G. : *A contribution to the theory of divergent sequences,* 1948, Vol. 80, pp. 167-190, Acta Math..

[9] Lorentz, G. G. : *Direct theorems on methods of summability,* 1949, Vol. 1, pp. 305-319, Can. J. of Math..

[10] Lorentz, G. G. : *Direct theorems on methods of summability II,* 1951, Vol. 3, pp. 236-256, Can. J. of Math..

[11] Lukacs, E. : *Characteristics functions,* 1970, 2nd edition, revised and enlarged, Hafner Publishing Co., NY.

[12] Meyer-König, W. : *Untersuchungen ueber einige verwandte Limitierungsverfahren,* 52, 1949, pp. 257-304, Math. Z.

[13] Peyerimhoff, A. : *Lectures on summability,* 1969, No. 107, Springer-Verlag Lecture Notes.

[14] Zeller, K. and Beekmann, W. : *Theorie der Limitierungsverfahren,* 1970, Ergaenzungen, Section 70, Math. Grenzgeb, 15, 2nd edition, Springer, Berlin.

In: Focus on Probability Theory
Editor: Louis R. Velle, pp. 43-72
ISBN: 1-59454-474-3

Chapter 3

SEMI-STABILITY OF SUMS AND MAXIMUMS IN SAMPLES OF RANDOM SIZE

S. Satheesh[1]
ARATHI, S. N. Park Road, Trichur – 680 004, India.
E. Sandhya[2]
Prajyoti Niketan College, Pudukkad
Trichur – 680 301, India.

Abstract

Semi-stable laws and max-semi-stable laws can be studied from two angles, (i) the semi-stability property of sums and maximums and (ii) their domains of partial attraction. In this chapter we discuss these laws from both the angles in samples of random size and they are referred to as φ-semi-stable and φ-max-semi-stable laws, where φ is a Laplace transform. For a given Laplace transform φ, a class of probability generating functions is presented. The integer-valued random variables corresponding to these probability generating functions satisfy certain conditions in the transfer theorems for sums and maximums. This facilitates the discussion of limit theorems for φ-semi-stable, φ-max-semi-stable and operator φ-semi-stable laws. Then we study the discrete φ-semi-stable laws and their attraction. φ-stable and φ-max-stable laws are discussed generalizing certain results on geometric stable and geometric max-stable laws. We give a method to construct max-self-decomposable laws, semi-self-decomposable laws and max-semi-self-decomposable laws. In the process we also address the question; why certain integer valued random variables do not induce the semi-stability property in the corresponding random sums/ maximums. Notable integer valued random variables in this category are Poisson and the negative binomial on {0,1,2, ...}. A necessary and sufficient condition for this is given.

[1] Email address: ssatheesh@sancharnet.in
[2] Email address: esandhya@hotmail.com

1 Introduction

Semi-stable and max-semi-stable laws have been envisaged as a generalization of the corresponding properties of stable and max-stable laws. While in stable/ max-stable laws the corresponding sum/ maximum of sample observations satisfies the stability property (invariance in distribution) with a change of scale $n^{-1/\alpha}$ for every sample of size n, semi-stable/ max-semi-stable laws satisfy the stability equation only for a fixed scale change $c^{-1/\alpha}$, $c>1$. Also, while stable/ max-stable laws have their domains of attraction/ max-attraction along $\{n\}$ the sequence of positive integers, semi-stable/ max-semi-stable laws have their domains of partial attraction along a sampling sequence $\{k_n\}$ satisfying, $\frac{k_n}{k_{n+1}} \to \frac{1}{c}, c \geq 1$. Some authors refer to this condition on the sampling sequence $\{k_n\}$ as a certain geometric growth and subsequently the attraction as geometric partial attraction. However, we reserve the terminology of geometric partial attraction exclusively for attraction when the sample size has a positive geometric distribution.

The development of semi-stable/ max-semi-stable laws is not just a technical curiosity. Semi-stable laws appear as the distributions of the increments of semi-stable Levy motions which models limits of random walks, fractional diffusion equations, see Meerschaert and Scheffler (2002), while their random-sum version has applications in continuous time random walks that are used in Physics to model a variety of phenomena connected with anomalous diffusion, see Meerschaert and Scheffler (2004), Kozubowski, *et al.* (2003). Max-semi-stable laws appear as the univariate marginal laws of a semi-self-similar extremal processes, see Pancheva (2000), Ben Alaya and Huillet (2002), while their random maximum version are related to the random time changed extremal processes, see Pancheva, *et al.* (2003). Random sums and random maximums have applications in Insurance and Hydrology, (Kaufman (2001)), Finance (Rachev (1993) and Mittnick and Rachev (1993)), Queuing (Makowski (2001)), Reliability (Cai and Kalashnikov (2000)), and Actuaries (Denuit, *et al.* (2002)). Key results in the discussion of attraction in the schemes of random sample sizes are the transfer theorems, Gnedenko and Korolev (1996). However, we have only a few examples of discrete laws that satisfy the requirements for invoking transfer theorems.

In this background this chapter is motivated by the need to discuss the semi-stability property and partial attraction when the sample size is random. Here we discuss a class of discrete laws that fit in to the scheme of samples of random size and transfer theorems so that we have a variety of options for the number of jumps by time $t>0$ in a continuous time random walk or the random time changed extremal processes model.

The chapter is organized as follows. In section.2 we discuss the semi-stability property of semi-stable/ max-semi-stable laws w.r.t the size of the sample. Probability generating functions (PGF) derived from Laplace transforms (LT) that are useful in the present context of samples of random size are discussed in section.3 including a class of PGFs useful in transfer theorems. The discussion extends the results of Gnedenko and Korolev (1996) and Satheesh (2002, 2004*a*). For any given LT φ we discuss φ-semi-stable and φ-max-semi-stable laws and their semi-stability properties w.r.t. the random size of the sample in section.4. This generalizes and extends the discussion in Satheesh *et al.* (2002). Here we also point out why certain integer-valued r.vs like Poisson or negative binomial do not induce semi-stability on any distribution. In section.5 we study the domains of partial attraction of φ-semi-stable and φ-max-semi-stable laws. Section.6 is devoted to a discussion on operator semi-stable laws and operator φ-semi-stable laws generalizing certain discussions in Kozubowski, *et al.* (2003) and Becker-Kern (2002*b*). Discrete semi-stable laws, discrete φ-semi-stable laws and their partial attraction are studied in section.7 extending certain results in Steutel and van-Harn (1979), Jayakumar (1995), Aly and Bouzar (2000) and Satheesh, *et al.* (2002). φ-stable and φ-max-stable laws are discussed in section.8 generalizing some of the results of Ramachandran (1997) and Rachev and Resnick (1991) on geometric/ max-geometric stable laws. Construction of extensions of self-decomposable laws is done in section.9 complementing the results on max-self-decomposable laws of Pancheva, semi-self-decomposable laws of Maejima and Naito (1998) and max-semi-self-decomposable laws of Becker-Kern (2001). Now we briefly describe semi-stable and max-semi-stable laws.

1.1 Semi-stable Laws

Levy introduced semi-stable laws in 1937, with characteristic functions (CF) f satisfying

$$f(t) = \{f(bt)\}^c \text{, for some } 0<|b|\leq 1<c. \tag{1}$$

In this setup there exists a unique $\alpha \in (0,2]$ such that $c|b|^\alpha = 1$ and we call this a semi-stable(c,b) law. Ramachandran and Rao (1968), Shimizu (1970), Pillai (1971), Kruglov (1972) and Mejzler (1973) have further studied these laws. In the sequel we will consider the case of $0<b<1$ only as the semi-stability property and domain of partial attraction can be more meaningfully discussed in this case. The case of $-1<b<0$ may be discussed on similar lines by replacing $f(bt)$ by $f(-bt)$ as suggested in Mohan, *et al.* (1993).

Since semi-stable laws are infinitely divisible (ID), setting $\psi(t) = -ln\{f(t)\}$ in (1) we have the class of continuous functions

$$\psi(t) = c\,\psi(bt) \text{, for all } t\in \mathbf{R} \text{ and some } 0<b<1<c \tag{2}$$

with $\psi(o) = 0$. They have been discussed in the context of CFs of semi-stable laws and regression equations in Kagan, *et al.* ((1973), p. 9, 163, 323, 324)), integrated Cauchy functional equations in Pillai and Anil (1996), a variation of it in approximating the gamma function in Dubec (1990) and in a study of the near-constancy phenomena in branching processes in Biggins and Bingham (1991). It has been proved that for (2) to hold there must exist a unique $\alpha>0$ such that $cb^\alpha=1$. α has to be restricted to (0,2] when $\phi(t) = exp\{-\psi(t)\}$ is a CF (that of a semi-stable law) and $\alpha\in(0,1]$ when $\phi(.)$ is a LT. Pillai and Anil (1996) have shown that the general solution of (2) is $\psi(t) = |t|^\alpha\, h(t)$ where $h(t)$ is a periodic function in $ln(t)$ with period $(-ln(b))$. $\psi(t) = \lambda|t|^\alpha$, $\lambda>0$ is clearly such a function.

Remark 1.1. In this context two problems are relevant. (i) Whether there are solutions to $\psi(t)$ other than $\psi(t) = \lambda|t|^\alpha$ or $\psi(s) = \lambda s^\alpha$, $\lambda>0$, so that $exp\{-\psi(.)\}$ is a CF or a LT (For $exp\{-\psi(s)\}$ to be a LT $\psi(s)$ should have completely monotone (CM) derivative), and (ii) conditions under which they reduce to $\lambda|t|^{\alpha}$ or λs^{α}. Regarding (i), when Pillai (1985) showed that

$$1/\{1+|t|^{\alpha}\{1-A\cos\{k\,ln(|t|)\}\}\},\ k=-2\pi/\,ln(b),\ 0<b<1 \text{ and } 0<A<1$$

is a Polya type CF, $exp\{-|t|^{\alpha}\{1-A\cos\{k\,ln(|t|)\}\}\}$ qualifies for the CF of a semi-stable law. In the case of LTs, CM solutions to $\xi(cs) = b\xi(s)$ were given by Dubec (1990), Biggins and Bingham (1991) and their results clearly show that there are functions $\xi(s)$ other than $s^{-\alpha}$ and under certain conditions $\xi(s)$ is close to a multiple of $s^{-\alpha}$. Bunge (1996) has used these results to describe LTs of semi-stable laws as $exp\{-\psi(s)\}$ where $\psi'(s) = \xi(s)$ and those of semi-Mittag-Leffler laws as $1/\{1+\psi(s)\} = \xi(s)/\{1+\xi(s)\}$. Jayakumar and Pillai (1993) (see also Satheesh and Nair (2002*a*) have given the example of the LT $1/\{1+\psi(s)\}$ with

$$\psi(s) = s^{\alpha}\{1-A\cos\{k\,ln(s)\}\},\ s>0,\ k=-2\pi/\,ln(b),\ 0<b<1 \text{ and } 0<A<1.$$

Thus definitions involving $\psi(.)$ lead to cases other than $\psi(t) = \lambda|t|^{\alpha}$ or $\psi(s) = \lambda s^{\alpha}$ so that (ii), the conditions under which they reduce to $\lambda|t|^{\alpha}$ or λs^{α} are also important. One such condition is when (2) is satisfied for two values of b, say b_1 and b_2 such that $ln(b_1)/\,ln(b_2)$ is irrational and especially if (2) is satisfied for all $0<b<1$, the case of stable laws. This is because if a periodic function has periods in irrational ratio then it must be a constant.

As limit laws, semi-stable laws are characterized as follows: Let $X_1, X_2, \ldots, X_n$ be independent and identically distributed (i.i.d) real r.vs with a common CF $g(t)$ and set $S_n = X_1 + X_2 + \ldots + X_n$, $n\geq 1$ integer. Let there exist sequences $\{a_n \in R\}$ and $\{b_n>0\}$ such that for some increasing sampling sequence $\{k_n\}$ satisfying

$$\frac{k_n}{k_{n+1}} \to \frac{1}{c},\ c\geq 1 \tag{3}$$

$$b_n^{-1} S_{k_n} - a_n \xrightarrow{d} Y. \tag{4}$$

Then Y is semi-stable and we say that the distribution of $\{X_i\}$ is in the domain of partial attraction of that of Y. Also (3) is equivalent to

$$\frac{b_n}{b_{n+1}} \to b. \tag{5}$$

Further in this setup also, as in the case of (1), there exists an $\alpha \in (0,2]$ such that $cb^{\alpha} = 1$ and the constants c and b identified in both the setups are the same. Notice that on iteration (1) is equivalent to:

$$f(t) = \{f(b^n t)\}^{c^n}. \tag{6}$$

One may choose $[c^n] = k_n$ and $b^{-n} = b_n$, Pillai (1971). In terms of CFs , (4) can be written as:

$$\{g(t/b_n)\, exp(-it\mu_n)\}^{k_n} \to f(t)\ ,\ \mu_n = a_n/k_n \tag{7}$$

Setting $g_n(t) = g(t/b_n)\, exp(-it\mu_n)$, (7) is equivalent to $k_n(1-g_n(t)) \to \psi(t)$ of (2). Still more generally, the condition $c_n(1-g_n(t)) \to \psi(t)$ for some real sequence $\{c_n>0\}$ satisfying

$$\frac{[c_n]}{[c_{n+1}]} \to \frac{1}{c}, c \geq 1 \tag{8}$$

is enough for the r.v Y to be semi-stable. Instead of (3) one may have the less restrictive condition, see Megyesi (2002),

$$\underset{k\to\infty}{Lt}\ \sup\,(k_n/k_{n+1}) = \frac{1}{c} \geq 0.$$

Some conditions for the r.v Y to be stable are, when:

(A) $c = 1$ in (3) or (8),
(B) $b = 1$ in (5),
(C) $k_n = n$ for all n,
(D) $[c_n] = n$ for all n,
(E) (2) is satisfied for two values of b, say b_1 and b_2 such that $ln(b_1)/\,ln(b_2)$ is irrational.

Remark.1.2 Notice that the class of stable laws obtained from semi-stable laws as limit laws of partial sums under the conditions A, B, C and D, mentioned above is larger than the one obtained under the condition E. This is because the condition E results in the CF $exp\{-\lambda|t|^{\alpha}\}$, $\lambda>0$ only, while as limit laws all the other well known forms of stable laws also result.

1.2 Max-semi-stable Laws

Grinevich and Pancheva were the first to describe max-semi-stable laws in 1992. From Megyesi (2002), max-semi-stable laws are of three types given by:

$$exp\{-e^{-x}\,\nu(x)\},\ x\in \boldsymbol{R}, \tag{9}$$

where $\nu(x)$ is a positive bounded periodic function with period $\beta>0$, and there exists a $c>1$ such that $ce^{-\beta} = 1$. Hence the period of $\nu(x)$ is $ln(c)$. This class is denoted as $\Lambda_{\beta,\,c}$. Again,

$$exp\{-x^{-\alpha}\,\nu(ln(x))\},\ x>0,\ \alpha>0, \tag{10}$$

where $\nu(x)$ is a positive bounded periodic function with period $ln(b)$, $b>1$, and there exists a $c>1$ such that $cb^{-\alpha} = 1$. This class is $\Phi_{b,c,\alpha}$. Finally,

$$exp\{-|x|^{\alpha}\nu(ln(|x|))\},\ x<0,\ \alpha>0, \tag{11}$$

where $\nu(x)$ is a positive bounded periodic function with period $|ln(b)|$, $b<1$, and there exists a $c>1$ such that $cb^{\alpha} = 1$. This class is $\Psi_{b,c,\alpha}$.

Remark 1.3 Notice that the d.fs in $\Phi_{b,c,\alpha}$ can be represented by the form $exp\{-\psi(x)\}$, where $\psi(x)$ satisfies $\psi(x) = c\,\psi(bx)$, $c>1$, $b>1$, and $\alpha>0$ satisfying $cb^{-\alpha} = 1$ and those in $\psi_{b,c,\alpha}$ by $exp\{-\psi(x)\}$, where $\psi(x)$ satisfies $\psi(x) = c\,\psi(bx)$, $c>1$, $b<1$, and $\alpha>0$ satisfying $cb^{\alpha} = 1$.

Examples for these laws are available in Pancheva (1994*a*, *b*). As limit laws max-semi-stable laws are characterized as follows, see Megyesi (2002). Let $X_1, X_2, \ldots, X_n$ be i.i.d real r.vs with a common d.f $F(x)$ and consider the normalized and centered maxima of them. If these normalized and centered maxima converges in distribution to a d.f G, that is;

$$(F(b_n x + a_n))^{k_n} \to G(x)\text{ , for all } x \in C_G\text{ , the set of continuity points of } G \tag{12}$$

where the increasing sub-sequence of positive integers $\{k_n\}$ satisfies,

$$\frac{k_{n+1}}{k_n} \to c \geq 1 \tag{13}$$

then the limiting d.f $G(x)$ is max-semi-stable and we say that F is in the domain of partial max-attraction of G. Also,

$$\frac{b_{n+1}}{b_n} \to 1,\ \frac{a_{n+1} - a_n}{b_n} \to \beta \text{ in the first case and}$$

$$\frac{b_{n+1}}{b_n} \to b,\ \frac{a_{n+1} - a_n}{b_n} \to 0 \text{ in the second and third cases.}$$

See Pancheva (1994*a*, *b*) and Megyesi (2002) for more on this. When $c = 1$ in (13), G is max-stable.

2 Semi-stability Property of Semi-stable/ Max-semi-stable Laws

Though the semi-stability equations (1), (9), (10) and (11) are well known the inherent semi-stability property of these laws with respect to the sample size is perhaps not well described. By the semi-stability property we mean: A distribution or its CF $f(t)$ has the semi-stability property in the summation scheme if there exists a $b>0$ and an integer $k>0$ such that

$$f(t) = \{f(bt)\}^{k} \text{ for all } t \in \mathbf{R}.$$

Similarly a distribution or its d.f $F(x)$ has the semi-stability property in the maximum scheme if there exists a $b>0$ and an integer $k>0$ such that

$$F(x) = \{F(bx)\}^k \text{ for all } x \text{ in the support of } F.$$

We may also see this as the property of scale invariance of sums/ maximums for a fixed number of i.i.d observations, that is, semi-stability in the strict sense. Now we have:

$$exp\{-\psi(t)\} = exp\{-c\,\psi(bt)\}$$

$$= \{exp\{-\psi(bt)\}\}^k \text{ when } c = k \text{, an integer.}$$

$$= \{exp\{-\psi(b^n t)\}\}^{k^n} \text{ on iteration. Hence;}$$

Theorem.2.1 A semi-stable(c,b) law is invariant under summation for a fixed sample of size $k \geq 1$ integer iff $c = k$, and the change of scale is $b = k^{-1/\alpha}$. The sample size can be k^n in which case the scale change is $b^n = k^{-n/\alpha}$.

Following a similar line of argument and by the property of the function $\psi(t)$ we have:

Theorem.2.2 A distribution is invariant under summation for samples of any size $k \geq 1$ integer iff it is stable.

Hence, if a semi stable law shows summation stability w.r.t samples of any size k then it must be stable. This conclusion is supported also by the fact that if this is true for every positive integer k, then there exists two integers say, k_1 and k_2 such that $ln(k_1)/ln(k_2)$ is irrational and consequently $\psi(t) = \lambda|t|^\alpha$ for a positive constant λ (condition E in section.1.1). Similarly for max-semi-stable laws we have the following two propositions.

Theorem.2.3 The maximum of k i.i.d observations of the type (10) with parameters b and k satisfy the semi-stability property with scale change $b = k^{1/\alpha}$.

Proof. Follows from $\{exp\{-(bx)^{-\alpha}\nu(ln(bx))\}\}^k = exp\{-x^{-\alpha}\nu(ln(x))\}$.

Theorem.2.4 The maximum of k i.i.d observations of the type (11) with parameters b and k satisfy the semi-stability property with scale change $b = k^{-1/\alpha}$.

Proof. Follows from $\{exp\{-|bx|^\alpha \nu(ln|bx|)\}\}^k = exp\{-|x|^\alpha \nu(ln|x|)\}$.

This is the scale invariance of (10) and (11). If these laws satisfy the property w.r.t two values of k say, k_1 and k_2 , such that $ln(k_1)/ ln(k_2)$ is irrational, then they become the corresponding max-stable laws of Gnedenko (condition E in section.1.1). The stability property of max-stable laws can thus be stated as:

Theorem.2.5 Max-stable laws are stable w.r.t taking the maximum of a random sample of any fixed size.

Theorem.2.6 The maximum of k i.i.d observations of the type (9) with parameters β and k satisfy another semi-stability property, the translation invariance property of (9).

Proof. Follows from

$$\{exp\{-e^{-(x+\beta)}\nu(x+\beta)\}\}^{k} = exp\{-e^{-x}\nu(x)\}.$$

Remark.2.1 Among the max-semi-stable laws, the classes $\Lambda_{\beta, k}$, $\Phi_{b, k, \alpha}$, $\Psi_{b, k, \alpha}$ fixed by the parameter $c = k$ alone are max-semi-stable when the sample size is k, and the change of location or scale is consequently fixed by the relation $ke^{-\beta} = 1$ or $cb^{-\alpha} = 1$ or $cb^{\alpha} = 1$ as the case may be.

3 Certain Useful Discrete Laws for Transfer Theorems

Here we discuss certain discrete laws and their PGFs derived from a LT that will be useful in our discussion of samples of random sizes. Limit theorems in schemes involving samples of random size can be formulated analogous to those in the classical setup using transfer theorems. If the sample size distribution N_θ satisfy (i) $\theta N_\theta \xrightarrow{d} U$ as $\theta \to 0$ where U is a non-negative r.v with LT φ, or (ii) $N_\theta \xrightarrow{p} \infty$ as $\theta \to 0$ in addition to (i), transfer theorems can be invoked, see Gnedenko and Korolev (1996). Under these two conditions we also need the r.vs $\{X_n\}$ under study to be independent of N_θ. Thus to model real life situations we need integer-valued r.vs that satisfy the conditions of the transfer theorems other than general results.

Sums and maximums and the corresponding limit results were discussed for samples of random sizes in the special cases of Poisson and positive geometric. While limit laws of Poisson sums had been part of the classical limit laws in the form of attraction and partial attraction Poisson maximums came in to picture in an analogous manner in max-ID laws in Balkema and Resnick (1977). The geometric law came in to the scheme of things in Renyi (1956) in the formulation of thinning of renewal processes (see also Sandhya (1991*b*)) and subsequently in the form of geometrically ID laws in Klebanov *et al.* (1984) and geometrically max-ID laws in Rachev and Resnick (1991) and Mohan (1998). The concept of attraction when the sample size distribution is positive geometric, namely geometric attraction, has been discussed in Sandhya (1991*a*) (see also Sandhya and Pillai (1999)), Mohan, *et al.* (1993) and max-geometric attraction in Rachev and Resnick (1991) and Mohan (1998). Later more general forms of random-ID (N-ID) laws were discussed in Sandhya (1991*a*, 1996), Gnedenko and Korolev (1996), Klebanov and Rachev (1996), Bunge (1996) and Kozubowski and Panorska (1996, 1998). Sandhya (1991*a*, 1996) defined N-ID laws as: A CF $f(t)$ is N-ID, where N is a positive integer valued r.v with PGF P_θ, if there exists a CF $f_\theta(t)$ such that

$$f(t) = P_\theta\{ f_\theta(t)\} \text{ for every } \theta \in \Theta. \tag{14}$$

Here N is assumed to have a finite mean and is independent of the distribution of $f_\theta(t)$ and Θ is the parameter space of θ. She also noticed that when $f(t)$ and $f_\theta(t)$ are of the same type (*ie*. when $f(t)$ is N-sum-stable) (14) is an Abel equation and discussed N-semi-stable laws as well. One drawback of this definition is that it is not constructive. However she did discuss two examples of non-geometric laws for N.

Gnedenko and Korolev (1996) and Klebanov and Rachev (1996) went further from (14), proving the de-Finetti analogue for N-sums *viz.* $f(t)$ is N-ID iff

$$f(t) = \underset{n\to\infty}{Lt}\ \varphi\{b_n(1-h_n(t))\} \tag{15}$$

where φ is a LT and $\{b_n\}$ are some positive constants and $\{h_n(t)\}$ are CFs. They then concluded that $f(t)$ is N-ID iff

$$f(t) = \varphi\{-\ln(\omega(t))\} \tag{16}$$

where $\omega(t)$ is the CF of an ID law. They first developed N-normal laws and then proceeded to define N-ID laws using the Poincare equation (Abel equation). This approach required that the family of PGFs $\{P_\theta : \theta \in \Theta\}$ formed a commutative semi-group with respect to the operation of convolution in addition to the existence of the mean of N. Here P_θ and φ (in (14), (15) and (16)) are related by

$$P_\theta(s) = \varphi\{\varphi^{-1}(s)/\theta\},\ \theta \in \Theta. \tag{17}$$

The discrete law corresponding to this PGF satisfies the requirement of Gnedenko's transfer theorem, Gnedenko and Korolev (1996). Bunge (1996) also arrived at (16) under the same assumptions but the arguments were based on Levy processes instead of proving the de-Finetti analogue enroute. The relation (17) holds true in this development also. Gnedenko and Korolev (1996) and Klebanov and Rachev (1996) also discuss attraction in this N-sum scheme.

All these authors also discuss N-stable laws while Klebanov and Rachev (1996) N-semi-stable laws as well, the requirement being that $\omega(t)$ in (16) must be the CF of a stable or a semi-stable law. For N-stable laws Kozubowski and Panorska (1996) have reached the same conclusion under the assumption that $N_\theta \xrightarrow{p} \infty$ as $\theta \downarrow 0$, without requiring that the family of PGFs of N_θ, $\{P_\theta : \theta \in \Theta\}$ forms a commutative semi-group.

Remark 3.1 Suppose $\varphi(s/\theta) = P_\theta\{\varphi(s)\}$ for every $\theta \in \Theta$. That is, the N_θ-sum of r.vs with LT $\varphi(s)$ is of the same type. This implies $P_\theta(s) = \varphi\{\varphi^{-1}(s)/\theta\}$, $\theta \in \Theta$. Thus (17) captures the structure of N-sum-stability. By corollary.4.6.1 in Gnedenko and Korolev (1996, p.141) this is equivalent to assuming that $\{P_\theta(s) : \theta \in \Theta\}$ form a commutative semi-group. See theorem.4.1*a* also.

Thus, as noticed by Satheesh (2002, 2004*a*), the development of N-ID laws based on the assumption that $\{P_\theta : \theta \in \Theta\}$ formed a commutative semi-group is not natural as it generalizes the notion of stability rather than infinite divisibility. From a methodological perspective also this is important as N-ID laws approximate N-sums and one cannot expect the distribution of N (sample size) to satisfy such an assumption. In fact the assumption rules out any P_θ having an atom at the origin. A case in point is when the distribution of N is Poisson or negative binomial and we do not have a corresponding N-ID law. On the other hand, when φ is gamma one expects the corresponding φ-ID law to approximate a negative binomial sum. But in this case P_θ corresponds to a Harris law see Satheesh (2004*a*). Another consequence of this assumption is that though the classical de-Finetti theorem describes an N-sum it does not fit in to this scheme of development here. Motivated by these Satheesh (2004*a*) discussed φ-ID laws circumventing the said drawbacks with respect to a wider class of PGFs, where φ is a LT. Later Satheesh (2002) and Satheesh and Sandhya (2003) extended the idea to φ-max-ID (φ-MID) laws.

Now we describe certain known results relating to LTs and PGFs that are useful here. Let $\varphi(s)$, $s>0$, is a LT of a non-negative probability law. Since LTs are decreasing in $(0,\infty)$ the inverse function φ^{-1} exists uniquely for a LT φ so that $\varphi\{\varphi^{-1}(s)\} = s$. Again, for any LT φ and a constant $0<\theta<1$, $\varphi(\theta s) = \varphi_\theta(s)$ is a LT.

Lemma 3.1 $\varphi\{\varphi^{-1}(s)/\theta\}$, $\theta>0$ is a PGF, Bunge (1996), Gnedenko and Korolev (1996) see also remark.3.1. In particular, setting $\varphi(s) = P_\theta\{\varphi(\theta s)\}$ we get $\varphi\{\varphi_\theta^{-1}(s)\}$, $0<\theta<1$, is a PGF.

Lemma 3.2 $\varphi\{(1-s)/\theta\}$, $0<s<1$, $\theta>0$ is a PGF, Satheesh (2001), Satheesh and Nair (2002*a*).

The following two lemmas are ramifications of Feller's proof (Feller, 1971, p.440) of Bernstein's theorem and were first discussed in Satheesh (2002) and also in Satheesh (2004*a*) and Satheesh and Sandhya (2003).

Lemma 3.3 $\wp_\varphi = \{P_\theta(s) = s^j\, \varphi\{(1-s^k)/\theta\}, 0<s<1, j\geq 0 \ \& \ k\geq 1$ integer and $\theta>0\}$ describes a class of PGFs for any given LT φ.

Lemma 3.4 Given a r.v U with LT φ, the integer valued r.vs N_θ with PGF P_θ in the class $\wp_\varphi$ described in lemma.3.3 satisfy $\theta N_\theta \xrightarrow{d} kU$ as $\theta \to 0$.

Satheesh (2004*b*) has used the class of PGFs in lemma.3.3 to discuss and give an example of a discrete ID law with gaps following the remark.9 in Bose, *et al.* (2002). Next we prove that $N_\theta \xrightarrow{p} \infty$ as $\theta \to 0$ for this class of PGFs. This facilitates the invoking of Szasz's (1972) converse to Gnedenko's transfer theorem.

Let $p_{n,k} = P\{N_n = k\}$, $k = 0, 1, 2, \ldots$. be the probability distribution corresponding to the sequence of r.vs $\{N_n, n\geq 1\}$. Let $P_n(s)$ denote the PGF of $\{N_n, n\geq 1\}$. Then;

$$N_n \xrightarrow{p} \infty \text{ as } n \to\infty \text{ is equivalent to } p_{n,k} \to 0 \text{ as } n \to\infty \text{ for every } k = 0, 1, 2, \ldots.$$

$$\text{Also, } \underset{n\to\infty}{Lt}\, P_n(s) = 0 \text{ for all } s\in(0,1) \Leftrightarrow \underset{n\to\infty}{Lt}\, p_{n,k} = 0, \text{ for every } k = 0, 1, 2, \ldots.$$

$$\Leftrightarrow N_n \xrightarrow{p} \infty \text{ as } n \to\infty.$$

Now consider the PGFs $\wp_\varphi = \{P_\theta(s) = s^j\, \varphi\{(1-s^k)/\theta\}, 0<s<1, j\geq 0 \ \& \ k\geq 1$ integer and $\theta>0\}$, where φ is the LT of an absolutely continuous distribution on $(0, \infty)$ so that $\varphi(\infty) = 0$. Also set $P_n(s) = s^j\, \varphi\{(1-s^k)/\theta_n\}$, where $\{\theta_n>0\}$ is a sequence of values of θ that decreases to 0 as $n \to\infty$. Then $\underset{n\to\infty}{Lt}\, P_n(s) = 0$ for all $s\in(0,1)$ as φ is the LT of an absolutely continuous law.

This proves:

Lemma 3.5 For every LT φ of an absolutely continuous law, we have a class of integer valued r.vs N_n with PGF P_n given by lemma.3.3 with $P_{\theta_n} = P_n$, satisfying $N_n \xrightarrow{p} \infty$ as $n \to \infty$.

4 φ-semi-stable and φ-max-semi-stable Laws and their Semi-stability Property

Satheesh (2002, 2004*a*) and Satheesh & Sandhya (2003) have introduced φ-infinitely divisible (φ-ID) and φ-MID (max-ID) laws for any given LT φ and studied them from various randomization schemes including random sums and maximums. They are defined as follows.

Definition 4.1 A CF $f(t)$ is φ-ID if there exists a LT φ and a CF $g(t)$ that is ID such that $f(t) = \varphi\{-ln(g(t))\}$, for all $t \in \mathbf{R}$.

Definition 4.2 A d.f F on $\mathbf{R}^d$, $d \geq 1$, is φ-MID if there exists a LT φ and a d.f G that is MID such that $F = \varphi\{-ln(G)\}$.

As semi-stable laws form a sub class of ID laws, φ-semi-stable and φ-max-semi-stable laws as defined below form a sub class of φ-ID and φ-MID laws:

Definition 4.3 A CF $f(t)$ is φ-semi-stable if there exists a LT φ and a CF $g(t)$ that is semi-stable such that $f(t) = \varphi\{-ln(g(t))\}$, for all $t \in \mathbf{R}$. When $g(t)$ is stable $f(t)$ is φ-stable.

Definition 4.4 A d.f F on $\mathbf{R}^d$, $d \geq 1$, is φ-max-semi-stable if there exists a LT φ and a d.f G that is max-semi-stable such that $F = \varphi\{-ln(G)\}$. When G is stable F is φ-max-stable.

Remark 4.1 A discussion of MID laws is relevant only in $\mathbf{R}^d$ since every d.f on $\mathbf{R}$ is MID. However max-semi-stable laws and max-stable laws can be discussed for d.fs on **R.**

Definition 4.4*a* φ-max-semi-stable laws corresponding to $\Phi_{b,\,c,\,\alpha}$ and $\psi_{b,\,c,\,\alpha}$ may be represented by the d.fs of the form $\varphi\{\psi(x)\}$, $x \in \mathbf{R}$, see remark.1.3.

Remark 4.2 These laws may also be thought of as mixtures of semi-stable/ max-semi-stable laws where the mixing law has LT φ (see Satheesh (2002)).

In terms of the PGF P_θ of the r.v N_θ, the N_θ-sum semi-stability of a CF $f(t)$ is:

$$P_\theta(f(\theta t)) = f(t) \text{ for some } \theta > 0 \text{ and all } t \in \mathbf{R}.$$

Semi-stability property by positive geometric sums was first discussed in Sandhya (1991*a*), as the property characterizing the semi-α-Laplace laws with CF $1/\{1+\psi(t)\}$ of Pillai (1985). See also Lin (1994, 1998). The result was generalized by Satheesh *et al.*(2002) where in they characterized the CF $1/\{1+\psi(t)\}^{1/k}$ with the semi-stability property by Harris-sums where the Harris(a,k) law has the PGF $s/\{a-(a-1)t^k\}^{1/k}$, $k>0$ integer and $a>1$. Similarly, the semi-stability property by geometric extremes were discussed in Pillai (1991), Pillai and Sandhya (1996), Satheesh (2001) and that by Harris extremes in Satheesh and Nair (2002*b*) and a generalization in Sreehari (1995). In what follows we use the notation $\varphi_\theta(s) = \varphi(\theta s)$.

Theorem 4.1 A φ-semi-stable law with parameters *c* and *b* is N_θ-sum semi-stable iff the PGF of *N* is $P_\theta(s) = \varphi\{\varphi_\theta^{-1}(s)\}$, $0<\theta<1$, the scale change being $0<b<1$ and $\theta = b^\alpha$.

Proof. Since $\varphi_\theta^{-1}\{\varphi_\theta(t)\} = t = \varphi^{-1}\{\varphi(t)\}$, we have: $\varphi_\theta^{-1}\{\varphi_\theta(\psi(t))\} = \varphi^{-1}\{\varphi(\psi(t))\}$

$$\Rightarrow \varphi_\theta^{-1}\{\varphi_\theta(c\psi(bt))\} = \varphi^{-1}\{\varphi(\psi(t))\}$$
$$\Rightarrow \varphi_\theta^{-1}\{\varphi(\theta c\psi(bt))\} = \varphi^{-1}\{\varphi(\psi(t))\}.$$

Since $cb^\alpha = 1$, when $\theta = b^\alpha$ we get

$$\varphi\varphi_\theta^{-1}\{\varphi(\psi(bt))\} = \varphi(\psi(t)).$$

Conversely, setting $\theta = b^\alpha$ we have:

$$\begin{aligned}\varphi\varphi_\theta^{-1}\{\varphi(\psi(bt))\} &= \varphi\varphi_\theta^{-1}\{\varphi(\theta c\psi(bt))\}\\ &= \varphi\varphi_\theta^{-1}\{\varphi(\theta\psi(t))\}\\ &= \varphi\varphi_\theta^{-1}\{\varphi_\theta(\psi(t))\}\\ &= \varphi\{\psi(t)\}, \text{ proving the theorem.}\end{aligned}$$

By corollary.4.6.1 in Gnedenko and Korolev (1996, p.141) the condition on the PGF P_θ in this theorem is possible iff $\{P_\theta, \theta \in \Theta\}$ form a commutative semi-group. Hence we have:

Theorem.4.1*a* A non-negative integer valued r.v $N(\theta)$ with PGF P_θ induces $N(\theta)$-sum semi-stability iff $\{P_\theta, \theta \in \Theta\}$ form a commutative semi-group.

Recalling remark.1.3 and invoking a similar line of argument shows that:

Theorem 4.2 φ-max-semi-stable laws in $\Phi_{b,c,\alpha}$, $\psi_{b,c,\alpha}$ are max-semi-stable w.r.t a sample size distribution N_θ iff the PGF of N_θ is $P_\theta(s) = \varphi\{\varphi_\theta^{-1}(s)\}$, $0<\theta<1$, the scale change being $0<b<1$ and $\theta = b^{-\alpha}$ or b^α as the case may be.

Theorem 4.2*a* A non-negative integer valued r.v $N(\theta)$ with PGF P_θ induces $N(\theta)$-max semi-stability iff $\{P_\theta, \theta \in \Theta\}$ form a commutative semi-group.

Theorem 4.3 φ-stable/ φ-max-stable laws are stable/ max-stable w.r.t the PGF $\varphi\{\varphi_\theta^{-1}(s)\}$ for any value of $0<\theta<1$.

Proof. Essentially follows from the fact that if a periodic function has periods in irrational ratio then it must be a constant.

Remark 4.3 Theorems 4.1*a*/ 4.2*a* also answer why certain integer valued random variables do not induce the semi-stability/ stability property in the corresponding random sums/ maximums. They give a necessary and sufficient condition for this. Thus Poisson and the negative binomial on {0,1,2, ...} do not induce the semi-stability/ stability property as their PGFs do not form a commutative semi-group. In particular, the condition in theorem.4.1 also states that $N(\theta)$ cannot have an atom at the origin. This is because at s=0, the RHS is zero and hence $P_\theta(o) = 0$. On the other hand, and more generally, it is also true that;

Theorem 4.4 When P{N=0}>0, absolutely continuous distributions cannot be represented as an N-sum. In particular they can never be N-sum stable if P{N = 0}>0.

Proof. If it were possible then there is a positive probability for the N-sum to be zero whereas the distribution cannot assume this value with positive probability being absolutely continuous. This is a contradiction, thus proving the assertion.

Now recall theorems 2.1 and 2.2. Feller (1971) had shown that if X is stable(α), then $X \stackrel{d}{=} n^{-1/\alpha}\{X_1+\ldots+X_n\}$ for every n. Here by theorem.4.1 we can see that when X is stable(α) a fixed sum alone imparts stability in the sum scheme and $c = n^{-1/\alpha}$ for every n, the number of components in the sum. Notice that for a semi-stable(k,b) law, k>1 integer fixes the number of components in the sum and it cannot vary, where as in the case of stable(α) laws the number of components n may vary, since a corresponding change in the scale parameter $c = n^{-1/\alpha}$ will guarantee stability.

Remark 4.4 Intuitively it thus appears that the condition

$$X \stackrel{d}{=} N^{-1/\alpha}\{X_1 + \ldots. + X_N\}$$

where N is a positive integer valued r.v , characterizes the stable(α) law. This has been proved in Ramachandran and Lau (1991, p.88).

5 Domain of Partial Attraction of φ-semi-stable and φ-max-semi-stable Laws

Recall definitions 4.3 and 4.4 here. One may notice that the term geometric attraction is also used for attraction along a subsequence $\{k_n\}$ of $\{n\}$ where $\{k_n\}$ shows a geometric growth, as in the case of semi-stable and max-semi-stable laws (see Megyesi (2002)). However we use this terminology exclusively for attraction of positive geometric-sums/ maximums.

Sandhya (1991a) (see also Sandhya and Pillai (1999)) had discussed attraction and partial attraction in positive geometric-sums for the first time while Rachev and Resnick (1991) and Mohan (1998) did it in positive geometric-maximums. Gnedenko and Korolev (1996) and Klebanov and Rachev (1996) have discussed attraction for N-sums when the r.v N has the PGF $\varphi\{\varphi^{-1}(s)/\theta\}$, θ>0. In the classical summation scheme a CF $h(t)$ belongs to the domain of attraction (DA) of the CF $\omega(t)$ if there exists a sequence of real constants b_n>0 and a_n such that

$$\text{as } n \to \infty,\ exp\{-n(1-h_n(t))\} \to exp\{-\psi(t)\} = \omega(t)\ \forall\ t \in \boldsymbol{R}.$$

where $h_n(t) = h(t/b_n)\, exp(-ita_n)$. A CF $h(t)$ belongs to the domain of partial attraction (DPA) of the CF $\omega(t)$ if there exists an increasing subsequence $\{k_n\}$ of positive integers such that

$$\text{as } n \to \infty,\ exp\{-k_n(1-h_n(t))\} \to exp\{-\psi(t)\} = \omega(t)\ \forall\ t\in \boldsymbol{R}.$$

From Pillai (1971) we have

Theorem 5.1 A semi-stable law belongs to its own domain of partial attraction.

Satheesh (2002, 2004*a*) has generalized the notions of geometric attraction and geometric max-attraction w.r.t sample size distributions N_θ , $\theta\in\Theta$ satisfying $\theta N_\theta \xrightarrow{d} U$ as $\theta \to 0$ where U is a positive r.v with LT φ, and N_θ are assumed to be independent of the distribution of the CF $h(t)$. This is formulated by considering $\{\theta_n\in\Theta\}$ such that as $n \to \infty$, $\theta \downarrow 0$ through $\{\theta_n\}$. The PGFs of N_θ will be denoted by $P_n = P_{\theta_n}$ corresponding to $\theta_n = 1/n$, where $\{n\}$ is the sequence of positive integers and $P_{k_n} = P_{\theta_n}$ corresponding to $\theta_n = 1/k_n$, a subsequence $\{k_n\}$ of $\{n\}$. Now with $h_n(t)$ as defined above we have the following description of φ-attraction (φ-A) and φ-max-attraction (φ-MA).

Definition 5.1 A CF $h(t)$ belongs to the domain of φ-attraction (Dφ-A) of the CF $f(t)$ if there exist sequences of real constants $b_n = b(\theta_n)>0$ & $a_n = a(\theta_n)$ such that $\underset{n\to\infty}{Lt}\ P_n\{h_n(t)\} = f(t)$, $\forall\ t\in \boldsymbol{R}$ and $h(t)$ is in the domain of partial φ-attraction (DPφ-A) of $f(t)$ if; $\underset{n\to\infty}{Lt}\ P_{k_n}\{h_n(t)\} = f(t)$, $\forall\ t\in \boldsymbol{R}$.

Definition 5.2 A d.f H in $\boldsymbol{R}^d$, $d \geq 2$ integer, belongs to the Dφ-MA of the d.f F with non-degenerate marginal distributions if there exists normalizing constants $b_{i,n} = b_i(\theta_n)>0$ and $a_{i,n} = a_i(\theta_n)$ such that with $H_n(y) = H(b_{i,n}y_i + a_{i,n}\ ,\ 1\leq i \leq d\ )$; $\underset{n\to\infty}{Lt}\ P_n\{H_n\} = F$ and H belongs to the DPφ-MA of the d.f F if; $\underset{n\to\infty}{Lt}\ P_{k_n}\{H_n\} = F$.

Remark 5.1 We now need a regularity assumption on the d.f G having φ as its LT, that is;

$$\int_0^\infty \omega_1^t\, dG(t) = \int_0^\infty \omega_2^t\, dG(t) \text{ implies } \omega_1 \equiv \omega_2 \text{ for any two CFs } \omega_1 \text{ and } \omega_2 \text{ that are ID,} \quad (18)$$

Gnedenko and Korolev (1996, p.108). This suggests that under the regularity assumption, the domain of attraction/ partial attraction is the same for the classical and random sum of i.i.d r.vs.

Theorem 5.2 Let $X_1, X_2, X_3, \ldots$ be i.i.d r.vs with CF $h(t)$ and N_θ an integer valued r.v corresponding to any PGF in $\wp_\varphi$ for a given LT φ. Then $h(t)$ is in the DPφA of the φ-ID law with CF $f = \varphi(\psi)$ iff $h(t)$ is in the DPA of the ID law with CF $e^{-\Psi}$.

Proof. Since $h(t)$ is in the DPφA of the φ-ID law with CF $f = \varphi(\psi)$ by the lemmata 3.4 and 3.5, the regularity assumption and Szasz's (1972) converse to Gnedenko's transfer theorem, $h(t)$ is in the DPA of the ID law with CF $e^{-\Psi}$. Converse follows by Gnedenko's transfer theorem and the proof is complete.

Satheesh and Sandhya (2003) gave the following necessary and sufficient condition for the CF $f(t) = \underset{\theta\to 0}{Lt}\, P_\theta\{h_\theta(t)\}$ to be φ-ID in terms of the CF h_θ, when the PGF of $N_\theta \in \wp_\varphi$.

Theorem 5.3 A CF $f = \underset{\theta\to 0}{Lt}\, \{h_\theta\}^j\, \varphi((1-\{h_\theta\}^k)/\theta)$, $j\geq 0$ & $k\geq 1$ integer, is φ-ID iff there exists a continuous function $\psi(t)$ such that $\forall\, t\in \boldsymbol{R}$, $(1-h_\theta(t))/\theta \to \psi(t)$, as $\theta \downarrow 0$ through a sequence $\{\theta_n\}$. In this case $f = \varphi(\psi)$.

Specializing these results to φ-semi-stable laws and with the above notations we have:

Theorem 5.4 A CF $f(t)$ is φ-semi-stable w.r.t any PGF in $\wp_\varphi$ for the given LT φ, iff there exists a CF $h(t)$, a sequence $\{\theta_n\}$ with $\{[1/\theta_n]\} = \{k_n\}$ satisfying $\frac{k_n}{k_{n+1}} \to \frac{1}{c}, c\geq 1$, and a continuous function $\psi(t)$ such that $\forall\, t\in \boldsymbol{R}$

$$(1-h_\theta(t))/\theta \to \psi(t), \text{ as } \theta \downarrow 0 \text{ through the sequence } \{\theta_n\}.$$

Theorem 5.5 A CF $f(t)$ has a nonempty DPφ-A w.r.t any PGF in $\wp_\varphi$ for a given LT φ, where $\{k_n\}$ satisfies $\frac{k_n}{k_{n+1}} \to \frac{1}{c}, c\geq 1$, iff it is φ-semi-stable.

Theorem 5.6 A φ-semi-stable law belongs to its own DPφ-A w.r.t any PGF in $\wp_\varphi$ for the given LT φ.

Theorem 5.7 The DPφ-A w.r.t any PGF in $\wp_\varphi$ for a given LT φ, of a φ-semi-stable law with CF $f(t) = \varphi\{\psi(t)\}$ coincides with the DPA of the semi-stable law with CF $e^{-\psi(t)}$.

Similarly for φ-MID laws Satheesh and Sandhya (2003) showed:

Theorem 5.8 A d.f $F = \underset{\theta\to 0}{Lt}\, \{H_\theta\}^j\, \varphi((1-\{H_\theta\}^k)/\theta)$, $j\geq 0$ & $k\geq 1$ integer, is φ-MID iff there exists a $\ell \in [-\infty,\infty)^d$ and an exponent measure μ concentrated on $[\ell,\infty]$ such that for $y\geq \ell$, $(1-H_\theta)/\theta \to \mu$, as $\theta \downarrow 0$ through a sequence $\{\theta_n\}$. In this case $F = \varphi(\mu)$.

Again as in the case of φ-semi-stable laws we have:

Theorem 5.9 A d.f F is φ-max-semi-stable w.r.t any PGF in $\wp_\varphi$ for the given LT φ, iff there exists a d.f H_θ, a sequence $\{\theta_n\}$ with $\{[1/\theta_n]\} = \{k_n\}$ satisfying $\frac{k_n}{k_{n+1}} \to \frac{1}{c}, c \geq 1$, and an exponent measure μ concentrated on $[\ell,\infty]$ such that for $y \geq \ell$

$$(1-H_\theta)/\theta \to \mu, \text{ as } \theta \downarrow 0 \text{ through the sequence } \{\theta_n\}.$$

Theorem 5.10 A d.f F has a nonempty DPφ-MA w.r.t any PGF in $\wp_\varphi$ for a given LT φ, where $\{k_n\}$ satisfies $\frac{k_n}{k_{n+1}} \to \frac{1}{c}, c \geq 1$, iff it is φ-max-semi-stable.

Theorem 5.11 The DPφ-MA of a φ-max-semi-stable law with d.f $F = \varphi(\mu)$, w.r.t any PGF in $\wp_\varphi$ for the given LT φ, coincides with the DPMA of the max-semi-stable law with d.f $e^{-\mu}$.

6 Operator Semi-stable/ φ-semi-stable Laws

Kozubowski, *et al.* (2003) discuss operator stable and operator ν-stable laws, while Becker-Kern (2002*a*, 2002*b*) studies operator semi-stable laws and their generalized domain of attraction with random sample sizes. Becker-Kern (2001, 2003) discusses the hemigroup structure of semi-stable, max-semi-stable laws and the corresponding self-decomposable laws. Lemma.3.5 enables us to discuss some of these theories with respect to the class of PGFs $\wp_\varphi$. The resultant operator semi-stable laws will be referred to as operator φ-semi-stable laws and they generalize the operator ν-stable laws of Kozubowski, *et al.* (2003) and operator φ-stable laws of Satheesh (2003).

Let $X, X_1, X_2, \ldots$be i.i.d random vectors on R^d with a common distribution μ and Y be a random vector on R^d whose distribution τ is full, that is, not supported on any lower dimensional hyperplane of R^d. τ is operator stable if there exists linear operators B_n on R^d and a sequence of points $a_n \in R^d$ such that:

$$B_n \sum_{i=1}^{n} (X_i - a_n) \xrightarrow{d} Y.$$

In terms of measures this is $B_n \mu^n * \varepsilon(s_n) \xrightarrow{w} \tau$, $\varepsilon(s_n)$ being the mass at the point $-nB_n a_n$.

Now let $U>0$ be a r.v with probability distribution ν and LT φ and suppose there exist integer valued r.vs $\{N_n\}$ independent of $\{X_i\}$ such that $N_n/n \xrightarrow{d} U$. Then Y_0 is operator ν-stable (Kozubowski, *et al.* (2003)) if there exist linear operators B_n on $\boldsymbol{R}^d$ and a sequence $a_n \in R^d$ such that:

$$B_n \sum_{i=1}^{N_n} (X_i - a_n) \xrightarrow{d} Y_0 \text{ or equivalently, } f(t) = \int_0^{\infty} \{\omega(t)\}^s \, d\nu(s);$$

where $f(t)$ & $\omega(t)$ are the CFs of Y_0 and Y respectively.

Using lemmata 3.3 and 3.4 Satheesh (2003) generalized the notion of operator ν-stable laws to operator φ-stable laws when the PGF of the r.vs $\{N_n\}$ belongs to $\wp_\varphi$. We may generalize this as follows.

Definition 6.1 Let $\{X_i\}$ be i.i.d random vectors on R^d, Y_0 a random vector on R^d with CF $f(t)$ whose distribution λ is full and $U>0$ be a r.v with probability distribution ν and LT φ. Then $f(t)$ (or its distribution or the random vector Y_0) is operator φ-stable if there exists integer valued r.vs $\{N_\theta\}$ independent of $\{X_i\}$ satisfying $\theta N_\theta \xrightarrow{d} U$ as $\theta \downarrow 0$, linear operators B_θ on R^d and a sequence $a_\theta \in R^d$ such that as $\theta \downarrow 0$ through $\{ \theta_n = 1/n\}$

$$B_\theta \sum_{i=1}^{N_\theta} (X_i - a_\theta) \xrightarrow{d} Y_0 \text{ or equivalently, } f(t) = \int_0^{\infty} \{\omega(t)\}^s \, d\nu(s) = \varphi\{-ln(\omega(t))\}.$$

$\omega(t)$ being the CF of Y, the operator stable law corresponding to $\{X_i\}$.

Theorem 6.1 Let $\{X_{\theta,j}\ ;\ \theta \in \Theta, j\geq 1\}$ be a sequence of i.i.d random vectors on $\boldsymbol{R}^d$, $d\geq 1$ integer, such that it has an operator limit law as $\theta \downarrow 0$ along $\{ \theta_n = 1/n \}$. Then the operator limit law of N_θ-sums of $\{X_{\theta,j}\}$ as $\theta \downarrow 0$ along $\{ \theta_n = 1/n \}$, where the PGF of N_θ is a member of $\wp_\varphi$, is necessarily operator φ-stable. Conversely, for any given LT φ; the operator φ-stable law can be obtained as the operator limit law of N_θ-sums of i.i.d random vectors as $\theta_n = 1/n \downarrow 0$ for each member of $\wp_\varphi$.

Remark 6.1 When N_n are positive geometric with mean n , U is standard exponential and Y_0 is operator exponential-stable with CF $f(t) = 1/(1-ln\{\omega(t)\})$. However the operator exponential-stable law can be obtained as the limit of random sums with respect to each integer-valued r.v whose PGF is in $\wp_\varphi$ where φ is the LT of an exponential law. Thus operator exponential-stable laws are conceptually more general than the operator geometric-stable laws of Kozubowski, *et al.* (2003).

Let $X, X_1, X_2, \ldots$.be i.i.d random vectors on R^d with a common distribution μ and Y be a random vector on R^d whose distribution τ is full. Then (Becker-Kern, 2002) τ is operator semi-stable if there exist linear operators B_n on R^d and a sequence of points $a_n \in R^d$ such that for some sampling sequence $\{k_n\}$ satisfying

$$\frac{k_n}{k_{n+1}} \to \frac{1}{c}, c \geq 1,$$

$$B_n \sum_{i=1}^{k_n} (X_i - a_n) \xrightarrow{d} Y.$$

In terms of measures this is $B_n \mu^{k_n} * \varepsilon(s_n) \xrightarrow{w} \tau$, $\varepsilon(s_n)$ being the mass at the point $-k_n B_n a_n$.

Definition 6.2 In this case we say that μ or X belongs to the generalized domain of partial attraction (GDPA) of the operator semi-stable law τ or Y.

Definition 6.3 Let $\{X_i\}$ be i.i.d random vectors on R^d, Y_0 be a random vector on R^d with CF $f(t)$ whose distribution λ is full and U>0 be a r.v with probability distribution ν and LT φ. Then $f(t)$ (or its distribution or the random vector Y_0) is operator φ-semi-stable if there exist integer valued r.vs $\{N_\theta\}$ independent of $\{X_i\}$ satisfying $\theta N_\theta \xrightarrow{d} U$ as $\theta \downarrow 0$, linear operators B_θ on R^d and a sequence of points $a_\theta \in R^d$ such that as $\theta \downarrow 0$ through $\{\theta_n = 1/k_n\}$

$$B_\theta \sum_{i=1}^{N_\theta} (X_i - a_\theta) \xrightarrow{d} Y_0 \text{ or equivalently, } f(t) = \int_0^\infty \{\omega(t)\}^s d\nu(s) = \varphi\{-ln(\omega(t))\}.$$

$\omega(t)$ being the CF of Y, the operator semi-stable law corresponding to $\{X_i\}$

Now conceiving the notion of generalized domain of partial φ-attraction (GDPφA) of an operator φ-semi-stable law on lines similar to definitions 5.1 and 6.2 and once again making use of lemma.3.5 and invoking the Szasz's (1972) converse to Gnedenko's transfer theorem under the regularity assumption (remark.5.1) on the distribution of the LT φ we have:

Theorem 6.2 Let $X, X_1, X_2, X_3, \ldots.$ be i.i.d random vectors in $\boldsymbol{R}^d$ with CF $h(t)$ and N_θ an integer valued r.v corresponding to any PGF in $\wp_\varphi$ for a given LT φ. Then X is in the GDPφA of the operator φ-semi-stable law with CF $f(t) = \varphi\{-ln(\omega(t))\}$ iff X is in the GDPA of the operator semi-stable law with CF $\omega(t)$.

Remark 6.2 One may define operator φ-semi-stable laws along the lines of definition.4.3 and then describe their GDPA instead of giving the definition.6.3. But the gain appears to be minimal.

7 Discrete Analogues of Semi-stable/ φ-semi-stable Laws

Stable family is a location-scale family of laws such that if r.vs X and Y each have distributions from that family then the distribution of their sum, i.e. $X+Y$, also belongs to the same family. Notice that in the definition of stable laws (Feller, 1971, p.170) the requirement is: X is stable if for each positive integer n there exists real constants b_n>0 and a_n such that with $\{X_i\}$ i.i.d as X and $S_n = X_1 + \ldots. + X_n$,

$$S_n \overset{d}{=} b_n X + a_n.$$

One can very well see that such a normalizing requirement would never be satisfied by a discrete law as the support immediately changes (Feller even leaves out degenerate laws from this class). Thus the definition is applicable only to continuous laws. So in the case of discrete laws the nature of normalization has to be different. Now consider positive stable laws with LT $exp\{-s^{\alpha}\}$, $\alpha \in (0,1)$, and notice that the LT $exp\{-s\}$ of the degenerate law also satisfies the stability requirement that the distributions must be invariant under summation. Invoking lemma.3.2 we can see that distributions with PGF $exp\{-\frac{1}{\theta}(1-s)^{\alpha}\}$, $\alpha \in (0,1]$, $\theta>0$ are also stable since the corresponding distributions are invariant under summation including the Poisson laws ($\alpha = 1$). This formulation is by Satheesh and Nair (2002*a*) or Satheesh, *et al.* (2002) and the approach is different from that of Steutel and van Harn (1979). Satheesh and Nair (2002*a*) has conceived the notion of discrete laws of the same type as:

Definition 7.1 Two PGFs $Q_1(s)$ and $Q_2(s)$ are of the same D-type iff

$$Q_1(1-s) = Q_2(1-bs), \text{ for all } 0<s<1 \text{ and some } 0<b<1.$$

From definition.7.1 we now formulate the notion of attraction as:

Definition 7.2 A discrete law with PGF $Q(s)$ is in the domain of attraction (DA) of another discrete law with PGF $R(s)$ if there exists a sequence $\{b_n>0\}$ such that

$$\lim_{n\to\infty} \{Q(1-b_n s)\}^n = R(1-s).$$

Remark 7.1 Setting $s = 1-u$ in this relation one can see that the formulation of attraction described above and in Steutel and van Harn (1979) are equivalent. Our approach also enables us to describe discrete semi-stable laws and partial attraction in the next stage.

A characteristic property of stable laws is that a distribution possesses a DA iff it is stable. For distributions on $[0,\infty)$ in terms of LTs, this is equivalent to (Feller,1971, p.449):

$$-n\, ln(\varphi(s/b_n)) \longrightarrow -ln(\varsigma(s)).$$

By lemma.3.2 we do have PGFs $\varphi\{(1-s)/b_n\}$ and $\varsigma(1-s)$, $0<s<1$ satisfying

$$-n\, ln(\varphi\{(1-s)/b_n\}) \longrightarrow -ln(\varsigma(1-s)).$$

Accordingly, $\varsigma(1-s) = exp\{-(1-s)^{\alpha}\}$, $0<\alpha\leq 1$ which is discrete stable. Hence:

Theorem 7.1 In the discrete case, discrete stable laws and they alone possess DA.

Theorem 7.2 A discrete stable law belongs to its own DA.

Thus stable laws are not necessarily absolutely continuous. It is the manifestation of the normalization that is different in the continuous and discrete supports. Discrete semi-stable laws were introduced in Satheesh and Nair (2002*a*) in analogy with the continuous case as follows.

Definition 7.3 A PGF $Q(s)$ is said to be discrete semi-stable if $Q(s) = exp\{-\psi(1-s)\}$, where $\psi(1-s) = c\psi\{b(1-s)\}$ for all $0<s<1$ and some $0<b<1<c$ satisfying $cb^{\alpha}=1$ for a unique $\alpha\in(0,1]$.

Stability properties of discrete geometric stable laws were studied in Jayakumar and Pillai (1992), Pillai and Jayakumar (1995), Aly and Bouzar (2000) and Satheesh and Nair (2002*a*). Semi-stability properties of discrete semi-stable laws, discrete geometric semi-stable laws and discrete generalized Linnik laws were considered in Jayakumar (1995), Satheesh and Nair (2002*a*) and Satheesh, *et al.*(2002). More general random summation schemes for discrete laws were discussed by Aly and Bouzar (2000) and Satheesh, *et al.*(2002). Now we discuss the semi-stability property and partial attraction in discrete semi-stable laws and their extensions and a generalization to discrete φ-semi-stable laws.

Definition 7.4 A discrete law with PGF $Q(s)$, is in the DPA of another discrete law with PGF $R(s)$ if there exists a sequence $\{b_n>0\}$ and a subsequence $\{k_n\}$ of $\{n\}$ such that

$$\lim_{n\to\infty}\{Q(1-b_n s)\}^{k_n} = R(1-s).$$

Definition 7.5 A PGF $Q(s)$ is discrete φ-semi-stable if $Q(s) = \varphi\{\psi(1-s)\}$, where $\psi(1-s)$ is as in definition.7.3 and φ is a LT. When $\psi(1-s) = (1-s)^{\alpha}$, $Q(s)$ is discrete φ-stable.

In analogy with the continuous case we now have:

Theorem 7.3 A discrete semi-stable law belongs to its own DPA.

Theorem 7.4 A PGF has a nonempty DPA where $\{k_n\}$ satisfies $\frac{k_n}{k_{n+1}}\to\frac{1}{c}, c\geq 1$, iff it is discrete semi-stable.

Theorem 7.5 As in theorem.4.1, a discrete φ-semi-stable law with parameters c and b is N-sum semi-stable iff the PGF of N is $P_\theta(s) = \varphi\{\varphi_\theta^{-1}(s)\}$, $0<\theta<1$, the scale change being $0<b<1$ and $\theta = b^{\alpha}$.

Remark 7.2 Aly and Bouzar (2000) have mentioned discrete N-ID and N-stable laws paralleling N-ID and N-stable laws of Gnedenko and Korolev (1996) and also the above result for discrete N-stable laws.

Now as in the built up of definition.5.1, let the sample size distributions N_θ, $\theta\in\Theta$ satisfy $\theta N_\theta \xrightarrow{d} U$ as $\theta\to 0$ where U is a positive r.v with LT φ, and N_θ are assumed to be independent of the distribution of the PGF $Q(s)$. To describe the sample size distribution consider the PGFs $P_n = P_{\theta_n}$ corresponding to $\theta_n = 1/n$, $\{n\}$ the sequence of positive integers and $P_{k_n} = P_{\theta_n}$ corresponding to $\theta_n = 1/k_n$, a subsequence $\{k_n\}$ of $\{n\}$.

Definition 7.6 A PGF $Q(s)$ belongs to the Dφ-A of the PGF $R(s)$ if there exists a sequence of real constants $b_n = b(\theta_n)$ such that $\underset{n\to\infty}{Lt}\ P_n\{Q(1-b_n s)\} = R(1-s)$, for all $s \in (0,1)$ and $Q(s)$ is in the DPφ-A of the PGF $R(s)$ if; $\underset{n\to\infty}{Lt}\ P_{k_n}\{Q(1-b_n s)\} = R(1-s)$, for all $s \in (0,1)$. Now we have:

Theorem 7.6 A PGF has a nonempty Dφ-A w.r.t any PGF in $\wp_\varphi$ for the given LT φ, iff it is discrete φ-stable.

Theorem 7.7 A PGF has a nonempty DPφ-A w.r.t any PGF in $\wp_\varphi$ for the given LT φ, where $\{k_n\}$ satisfies $\dfrac{k_n}{k_{n+1}} \to \dfrac{1}{c}, c \geq 1$, iff it is discrete φ-semi-stable.

Theorem 7.8 A discrete φ-stable law belongs to its own Dφ-A w.r.t any PGF in $\wp_\varphi$ for the given LT φ.

Theorem 7.9 A discrete φ-semi-stable law belongs to its own DPφ-A w.r.t any PGF in $\wp_\varphi$ for the given LT φ.

Theorem 7.10 The Dφ-A w.r.t any PGF in $\wp_\varphi$ for a given LT φ, of a discrete φ-stable law with PGF $Q(s) = \varphi\{(1-s)^\alpha\}$ coincides with the DA of the discrete stable law with PGF $exp\{-(1-s)^\alpha\}$.

Theorem 7.11 The DPφ-A w.r.t any PGF in $\wp_\varphi$ for a given LT φ, of a discrete φ-semi-stable law with PGF $Q(s) = \varphi\{\psi(1-s)\}$ coincides with the DPA of the discrete semi-stable law with PGF $exp\{-\psi(1-s)\}$.

Thus we can discuss the DA of discrete stable (Dφ-A of discrete φ-stable) and the DPA of discrete semi-stable (DPφ-A of discrete φ-semi-stable) laws parallel to the continuous case.

8 φ-stable/ φ-max-stable Laws

Here we generalize two results of Ramachandran (1997) on geometric-stable laws and stable processes and another one by Rachev and Resnick (1991) on geometric-max-stable laws and processes. Recall that for a LT φ, a CF $\varphi(\psi_\alpha)$ is φ-stable iff $exp\{-\psi_\alpha\}$ is a stable(α) CF.

Theorem 8.1 A φ-stable r.v has the representation

$$XZ^{1/\alpha} + \mu(Z-Z^{1/\alpha}),\ \alpha\neq 1$$
$$XZ + (2\beta\lambda/\pi)\, Z \log Z,\ \alpha=1$$

Where X is stable(α) and Z is a r.v independent of X having LT φ and d.f G.

Proof. Proceeding as in Ramachandran (1997, p.303) the CF of the above r.v is:

$$\int_0^\infty e^{-t\psi_\alpha(u)}dG(t) = \varphi(\psi_\alpha(u)), \text{ completing the proof.}$$

Theorem 8.2 Let T be a r.v independent of $\{X(t)\}$ with LT φ and d.f G. Then a necessary and sufficient condition for $\{X(t)\}$ to be a stable(α) process is that the r.v $\left(\int_0^T t dX(t)\right)/T$ have the same distribution as $X(\gamma T) + \delta T$, for some $\gamma>0$ and real δ. Then necessarily $\frac{1}{3}<\gamma<1$. The common d.f of these two r.vs is then φ-stable with the same exponent as the process, the precise relationship being given by (19).

Proof. Let $e^{-\psi}$ be the CF of $X(1)$. Then proceeding as in Ramachandran (1997, p.306),

$$\mathrm{E}\, e^{iu(\int_0^T t dX(t)/T)} = \int_0^\infty \exp[-t\int_0^1 \psi(uv)dv]dG(t)$$
$$= \varphi(\int_0^1 \psi(uv)dv).$$

$$\text{Also, } \mathrm{E}\, e^{iu[X(\gamma T)+\delta T]} = \varphi\{-i\delta u + \gamma\psi(u)\}. \tag{19}$$

Our identical distribution assumption thus implies:

$$\int_0^1 \psi(uv)dv = -i\delta u + \gamma\psi(u) \tag{20}$$

where $e^{-\psi}$ is a CF. Equation (20) characterizes the stable(α) CF with exponent $\alpha = (1/\gamma) - 1$ and hence $\frac{1}{3}<\gamma<1$. Consequently the common distribution of the two r.vs considered is φ-stable with the same exponent as the process itself. That completes the proof of the proposition.

Theorem 8.3 For a d.f F on $\boldsymbol{R}^d$, the following are equivalent.

i. F is φ-max-stable
ii. $exp\{-\varphi^{-1}(F)\}$ is max-stable
iii. There exists a $\ell\in[-\infty,\infty)^d$ and an exponent measure μ concentrated on $[\ell,\infty]$ such that for $y\geq\ell$, $F(y) = \varphi\{\mu([\ell,y]^c\}$

iv. There exists a multivariate extremal process $\{Y(t), t>0\}$ governed by a max-stable law and an independent r.v Z with d.f F_Z and LT φ such that $F(y) = \mathrm{P}\{Y(Z) \le y\}$.

Proof. (i) $\Rightarrow$ (ii) F is φ-max-stable implies $F = \varphi\{-\ln G\}$ where G is max-stable

$$\Rightarrow exp\{-\varphi^{-1}(F)\} = G \text{ is max-stable.}$$

(ii) $\Rightarrow$ (iii) By (ii) $exp\{-\varphi^{-1}(F)\} = G = exp\{-\mu([\ell,y]^c)\}$.

$$\text{Hence; } \varphi^{-1}(F) = \mu([\ell,y]^c) \text{ or } F(y) = \varphi\{\mu([\ell,y]^c\}.$$

(iii) $\Rightarrow$ (iv) By (iii) we have an exponent measure μ and let $\{Y(t), t>0\}$ be the extremal process governed by a max-stable law with exponent measure μ.

That is; $\mathrm{P}\{Y(t) \le y\} = exp\{-t\mu([\ell,y]^c)\}$. Hence,

$$\mathrm{P}\{Y(Z) \le y\} = \int_0^\infty exp\{-t\mu([\ell,y]^c)\}\, dF_Z(t) = \varphi\{\mu([\ell,y]^c\} = F(y).$$

(iv) $\Rightarrow$ (i) This is now obvious. Thus the proof is complete.

This result generalizes Proposition.3.2 in Rachev and Resnick (1991) when N_θ is geometric. One my reformulate this result for φ-max-semi-stable laws.

9 Extensions of Self-decomposable Laws

Definition 9.1 A CF f is self-decomposable if for every $0<c<1$ there is a CF g_c such that

$$f(t) = f(ct)\, g_c(t) \text{ for all } t \in \mathrm{R}.$$

For a recent review of self-decomposable laws with examples and Levy measures see Jurek (1997, 2001) and Satheesh (2003). Sandhya and Satheesh (1996), Ramachandran (1997) and Satheesh, *et al.* (2002) have discussed the relation between random sum stability and self-decomposability. From the angle of φ-mixtures of ID laws Satheesh (2002) has generalized Theorem.2.1 (part (iii), (iv) and (v)) of Ramachandran (1997) and Theorems 2.3, 2.4 and 2.5 on the self-decomposability of generalized Linnik laws of Erdogan and Ostrowski (1998) in Satheesh, *et al.* (2002). Satheesh (2002) gave a method to construct self-decomposable laws as:

Theorem 9.1 A φ-mixture of strictly stable laws is self-decomposable if φ is self-decomposable.

Different authors have studied extensions of the concept of self-decomposable laws. Both semi-stability and self-decomposability are extensions of the notion of stability in the sense of limit laws. While in semi-stability we are considering limit laws of partial sums of i.i.d r.vs where the sampling sequence $\{k_n\}$ satisfies the condition (3) in self-decomposability we are considering limit laws of partial sums of independent r.vs and the sampling sequence is $\{n\}$. Now when we consider limit laws of partial sums of independent r.vs and the sampling sequence $\{k_n\}$ satisfies the condition (3) we have semi-self-decomposable laws. Similarly one may consider the limit laws of maximums. Maejima and Naito (1998) studied semi-self-decomposable laws, Becker-Kern (2001) studied max-semi-self-decomposable laws extending the terminology of max-self-decomposable laws of Pancheva (1994*b*). Next we give methods to construct the above extensions of self-decomposable laws.

Definition 9.2 A non-degenerate d.f F is max-self-decomposable if for every $c\in(0,1)\cup(1,\infty)$ there is a non-degenerate d.f H_c such that

$$F(x) = F(cx)\, H_c(x) \text{ for all } x\in \mathbf{R}.$$

Definition 9.3 A CF f is semi-self-decomposable if for some $0<c<1$ there is a CF g_c such that

$$f(t) = f(ct)\, g_c(t) \text{ for all } t\in \mathbf{R}.$$

Definition 9.4 A non-degenerate d.f F is max-semi-self-decomposable if for some $c\in(0,1)\cup(1,\infty)$ there is a non-degenerate d.f H_c such that

$$F(x) = F(cx)\, H_c(x) \text{ for all } x\in \mathbf{R}.$$

Theorem 9.2 φ-max-stable laws corresponding to (10) and (11) are max-self-decomposable if φ is self-decomposable.

Proof. If φ is self-decomposable then $\varphi(s) = \varphi(cs)\, \varphi_c(s)$, $s>0$, for every $0<c<1$.

Now the d.fs of φ-max-stable laws corresponding to (11) can be written as:

$$\varphi\{-|x|^{\alpha}\} = \varphi\{-c|x|^{\alpha}\}\, \varphi_c\{-|x|^{\alpha}\} \text{ for every } 0<c<1.$$

Similar is the case with (10). That completes the proof.

Example 9.1 The extended log-logistic law of Voorn (1987) with d.f $F(x) = \{1+x^{-\alpha}\}^{-\beta}$, $x>0$, $\alpha>0$, $\beta>0$, is N-max stable iff $\beta = 1/k$, $k>0$ integer, and N is Harris$(c^{-\alpha},k)$, Satheesh and Nair (2002*b*). Since the log-logistic law is a gamma mixture of max-stable laws corresponding to (10) and the gamma$(1,\beta)$ law is self-decomposable, the log-logistic law is max-self-decomposable.

Theorem 9.3 φ-semi-stable laws are semi-self-decomposable if φ is self-decomposable.

Proof. If φ is self-decomposable then the CF of φ-semi-stable laws can be written as:

$$\varphi\{\psi(t)\} = \varphi\{c\,\psi(t)\}\, \varphi_c\{\psi(t)\} \text{ for every } 0<c<1 \text{ and for all } t\in \mathbf{R},$$

where $\psi(t) = a\,\psi(bt)$, $a>1$, $b<1$, and $\alpha>0$ satisfying $ab^{\alpha} = 1$. When $c = b^{\alpha}$ we have:

$\varphi\{\psi(t)\} = \varphi\{\psi(bt)\}\ \varphi_b\{\psi(t)\}$ for some $0<b<1$, completing the proof.

Example 9.2 The generalized semi-α-Laplace law with CF $f(t) = \{1+\psi(t)\}^{-\beta}$, $\psi(t) = c\psi(bt)$ as in (2) and $\beta>0$, is N-sum stable iff N is Harris(c,k), $\beta = 1/k$ which follows from throrem.4.1. Since the gamma$(1,\beta)$ law is self-decomposable and the generalized semi-α-Laplace law is a gamma mixture of semi-stable laws, the generalized semi-α-Laplace law is semi-self-decomposable.

Theorem 9.4 φ-max-semi-stable laws corresponding to (10) and (11) are max-semi-self-decomposable if φ is self-decomposable.

Proof. Follows along the same lines as the proofs of theorems 9.2 and 9.3.

Example 9.3 Considering a gamma mixture of max-semi-stable laws corresponding to (10) one has the d.f $F(x) = \{1+\psi(x)\}^{-\beta}$, $x>0$, $\psi(x) = c\psi(bx)$, $\beta>0$. This d.f is max-semi-self-decomposable since the gamma$(1,\beta)$ law is self-decomposable.

10 Conclusion

Thus in this chapter we have introduced φ-semi-stable, φ-max-semi-stable and operator φ-semi-stable laws and discussed their semi-stability property and domains of partial attraction when the sample size is random. A necessary and sufficient condition to be satisfied by an integer-valued r.v N for inducing the semi-stability property in the N-sum or N-maximum scheme is also given. This explains why Poisson and negative binomial r.vs do not induce semi-stability. Classes of PGFs that facilitates the study of semi-stability property as well as random limit theorems by invoking transfer theorems are discussed. Discrete analogues of semi-stable and φ-semi-stable laws are introduced and studied. Constructions of certain extensions of self-decomposable laws are discussed as a by-product of φ-semi-stable and φ-max-semi-stable laws. φ-stable and φ-max-stable laws are also studied in the sequel.

We know that there exist processes with stationary and independent increments corresponding to each semi-stable and stable law, see Meerschaert and Scheffler (2002). The randomization introduced by way of φ and the resulting limit laws make them useful in modeling anomalous diffusions and continuous time random walks. Continuous time random walks were introduced in Montroll and Weiss (1965) and are now used in Physics to model anomalous diffusion and relaxation, see Klafter, *et al.* (1987), Scher and Lax (1973), Uchaikin and Zolotarev (1999) and Kotuluski and Weron (1996). Meerschaert and Scheffler (2004) and Kozubowski, *et al.* (2003) discuss limit theorems for continuous time random walks and show that the limit is either operator ν-stable or belongs to the generalized domain of normal attraction of an operator stable law. Here the random sample size corresponds to the number of jumps by time $t\geq 0$. Now by lemma.3.5, we have a class of discrete laws for a given LT φ that can model this number of jumps and hence our operator φ-stable laws are useful in a variety of such situations.

By lemma.3.5 we have a family of r.vs $\{N_\theta,\ \theta>0\}$ satisfying $N_\theta \xrightarrow{p} \infty$ as $\theta \to 0$. Hence we can invoke the results of Becker-Kern (2002*b*), his stochastic summability method, mantissa distribution and study properties of randomized max-semi-stable and randomized operator semi-stable laws w.r.t this class of positive integer valued r.vs. Becker-Kern (2002*a*) studies operator semi-stable limit laws with random sample size under the stronger condition that the r.vs $\{T_n\}$ describing the sample size distribution satisfies $T_n/k_n \xrightarrow{d} D$ for some r.v $D>0$. In this setup one need not assume that $\{T_n\}$ is independent of $\{X_i\}$.

It is known that max-semi-stable and max-stable laws appear as univariate marginal distributions of semi-self-similar and self-similar extremal processes, Pancheva (1998, 2000), Ben Alaya and Huillet (2003). The φ versions of max-semi-stable and max-stable laws discussed here make them useful in discussing the corresponding random time changed extremal processes of Pancheva *et al.*(2003). Attraction in N-max and N-min schemes of discrete laws are yet to be studied though Satheesh and Nair (2002*b*) had discussed N-max and N-min stability properties for discrete laws.

Gaver and Lewis (1980) showed how self-decomposable laws can be useful in modeling AR(1) processes. When we consider maxification processes along these lines one has to use max-self-decomposable laws for their modeling. Construction of different extensions of self-decomposable laws discussed in section.9 has a role to play here. Recently the authors of this chapter have (i) characterized semi-selfdecomposable/ max-semi-selfdecomposable laws as the marginal distributions of AR(1)/ max-AR(1) models, (ii) discussed semi-self-decomposable laws on non-negative integers and integer valued AR(1) models, (iii) showed that semi-stable/ max-semi-stable laws are semi-self-decomposable/ max-semi-self-decomposable and (iv) developed AR(1) models with semi-stable/ max-semi-stable marginals. See Satheesh and Sandhya (2006) for an update on this. Pancheva (1994*a*, *b*) have considered max-self-decomposability under non-linear and monotone normalizations.

Another interesting line of development in semi-stable/ max-semi-stable laws is the almost sure limit theorems for sums and maxima. It has been shown that the sums/ maxima from the DPA/ DPMA of a semi-stable/ max-semi-stable law satisfy almost sure limit theorems along the $\{n\}$ though weak convergence takes place only along a subsequence $\{k_n\}$. Results on this kind of merging are the focus in Megyesi (2002) and Berkes, *et al.* (2002).

References

[1] Aly, A. A.; Bouzar, N. On geometric infinite divisibility and stability. *Ann. Inst. Statist. Math.* 2000, 52, 790–799.

[2] Balkema, A. A.; Resnick, S. Max-infinite divisibility. *J. Appl. Probab.* 1977, 14, 309–319.

[3] Becker-Kern, P. Max-semistable hemigroups: structure, domain of attraction and limit theorems with random sample size. *Probab. Math. Statist.* 2001, 21, 441–465.

[4] Becker-Kern, P. Limit theorems with random sample size for generalized domains of semistable attraction. *J. Math. Sci.* 2002 (*a*), 111, 3820–3829.

[5] Becker-Kern, P. Stochastic summablity methods for domains of normal attraction of semistable laws, preprint, 2002 (*b*).

[6] Becker-Kern, P. Stable and semistable hemigroups: Domains of attraction and self-decomposability. *J. Theor. Probab.* 2003, 16, 573–598.

[7] Ben Alaya, M. ; Huillet, T. On multiscaling max-semistable distributions. preprint, 2003.

[8] Berkes, I; Csaki, E; Csorgo, S. ; Megyesi, Z. Almost sure limit theorems for sums and maxima from the domain of geometric partial attraction of semistable laws. 2002, preprint, to appear in *Proceedings of the Fourth Hungarian Colloquim on Limit Theorems in Probability and Statistics*, Berkes, *et al.* (Ed.s) *J. Bolyai Math. Soc.* Budapest.

[9] Bose, A; Dasgupta, A. ; Rubin, H. A contemporary review and bibliography of infinitely divisible distributions and processes. *Sankhya-A*. 2002, 64, 763–819.

[10] Biggins, J. D. ; Bingham, N.H. Near-constancy phenomena in branching processes. *Math. Proc. Camb. Phil. Soc.* 1991, 110, 545–558.

[11] Bunge, J. Composition semi-groups and random stability. *Ann. Probab.* 1996, 24, 1476–1489.

[12] Cai, I. ; Kalashnikov, V. NWU property of a class of random sums, *J. Appl. Probab.* 2000, 37, 283–289.

[13] Denuit, M; Geneset, C. ; Marceau, E. Criteria for the stochastic ordering of random sums, with Acturial applications. *Scand. J. Statist.* 2002, 3-16.

[14] Dubec, S. An approximation of the gamma function. *J. Math. Anal. Appl.* 1990, 146, 461–468.

[15] Erdogan, M. B. ; Ostrovskii, I. V. Analytic and asymptotic properties of generalized Linnik probability densities. *J. Math. Anal. Appl.* 1998, 217, 555-578.

[16] Feller, W. *An Introduction to Probability Theory and Its Applications*, Wiley, New York, 1971; Vol.2.

[17] Gaver, D. P. ; Lewis, P. A. W. First-order auto regressive gamma sequences and point processes. *Adv. Appl. Probab.* 1980, 12, 727–745.

[18] Gnedenko, B. V. ; Korolev, V. Yu. *Random Summation - Limit Theorems and Applications*, CRC Press, Boca Raton, 1996.

[19] Jayakumar, K. The stationary solution of a first order integer valued autoregressive process. *STATISTICA*. 1995, anno LV, 221–228.

[20] Jayakumar, K. ; Pillai, R. N. On class-L distributions. *J. Ind. Statist. Assoc.* 1992, 30, 103-108.

[21] Jayakumar, K. ; Pillai, R. N. First order autoregressive Mittag-Leffler process. *J. Appl. Probab.* 1993, 30, 462–466.

[22] Jurek, Z. J. Self-decomposability: an exception or a rule? *Ann. Univ. Mariae Curie Sklodowska Lubin Polonia*. 1997, L1.1, Section.A, 93–106.

[23] Jurek, Z. J. Remarks on self-decomposability and new examples. *Demonstratio Math.* 2001, XXXIV, 241-250.

[24] Kagan, A. M; Linnik, Yu. V. ; Rao, C. R. *Characterization Problems in Mathematical Statistics*, Wiley, New York, 1973.

[25] Kaufman, E. *Statistical Analysis of Extrme Values, From Insurance, Finance, Hydrology and Other Fields*, Extended 2nd Edition, Birkhauser, 2001.

[26] Klafter, J.; Blumen, A.; Shlesinger, M. F. Stochastic pathways to anomalous diffusion. *Phys. Rev. A*. 1987, 35, 3081–3085.

[27] Klebanov, L. B. ; Rachev, S. T. Sums of a random number of random variables and their approximations with ν-accompanying infinitely divisible laws. *Serdica Math. J.* 1996, 22, 471–496.

[28] Klebanov, L. B; Maniya, G. M. ; Melamed, I. A. A problem of Zolotarev and analogues of infinitely divisible and stable distributions in the scheme of summing a random number of random variables. *Theor. Probab. Appl.* 1984, 29, 791–794.

[29] Kotulski, M. ; Weron, K. Random walk approach to relaxation in disordered systems. *Proceedings of Athens Conference on Applied Probability and Time Series Analysis, Lecture Notes in Statistics.* 114, Springer, New York, 1996; Vol.1, 379–388..

[30] Kozubowskii, T, J. ; Panorska, A. K. On moments and tail behavior of ν-stable random variables. *Statist. Probab. Lett.* 1996, 29, 307-315.

[31] Kozubowskii, T. J. ; Panorska, A. K. Weak limits for multivariate random sums. *J. Multi. Anal.* 1998, 67, 398-413.

[32] Kozubowski, T. J; Meerschacrt, M. M ; Scheffler, H. P. The operator ν-stable laws. *Publ. Math. Debrecen.* 2003, 63, 569–585.

[33] Kruglov, V. M. On the extension of the class of stable distributions. *Theor. Probab. Appl.* 1972, 17, 685–694.

[34] Lin, G. D. Characterizations of the Laplace and related distributions via geometric compound. *Sankhya–A.* 1994, 56, 1–9.

[35] Lin, G. D. A note on Linnik distributions. *J. Math. Anal. Appl.* 1998, 217, 701–706.

[36] Maejima, M ; Naito, Y. Semi-selfdecomposable distributions and a new class of limit theorems. *Probab. Theor. Rel. Fields.* 1998, 112, 13–31.

[37] Makowski, A. M. On a random sum formula for the busy period of the M/G/∞ queue with applications. Technical Research Report-CSHCN TR 2001-4, *University of Maryland*, 2001.

[38] Meerschacrt, M. M. ; Scheffler, H. P. Semistable Levy motion. *Fractional Calculus Appl. Anal.* 2002, 5, 27–54.

[39] Meerschacrt, M. M. ; Scheffler, H. P. Limit theorems for continuous time random walks with infinite mean waiting times, preprint, 2004.

[40] Megyesi, Z. Domains of geometric partial attraction of max-semistable laws: structure, merge and almost sure limit theorems. *J. Theor. Probab.* 2002, 15, 973–1005.

[41] Mejzler, D. On a certain class of infinitely divisible distributions. *Israel J. Math.* 1973, 16, 1-19.

[42] Mittnick, S. ; Rachev, S. T. Modeling asset returns with alternative stable distributions. *Econometric Rev.* 1993, 12, 161–330.

[43] Mohan, N. R. On geometrically max infinitely divisible laws. *J. Ind. Statist. Assoc.* 1998, 36, 1-12.

[44] Mohan, N. R; Vasudeva, R. ; Hebbar, H. V. On geometrically infinitely divisible laws and geometric domains of attraction. *Sankhya-A.* 1993, 55, 171–179.

[45] Montroll, E. W. ; Weiss, G. H. Random walks on lattices II. *J. Math. Phys.* 1965, 6, 167-181.

[46] Pancheva, E. Extreme value limit theory with non-linear normalization. in *Extreme Value Theory and Applications*. J. Galambos, *et al.* (Eds.), Kluwer Academic Publishers, 1994 (*a*), 305–318.

[47] Pancheva, E. Multivariate extreme value limit distributions under monotone normalization. in *Stability Problems for Stochastic Models.* V. M. Zolotarev, *et al.* (Eds.), TVP/ VSP, Moscow, 1994 (*b*), 179–196.

[48] Pancheva, E. Self-similar extremal processes. *J. Math. Sci.* 1998, 92, 3911–3920.

[49] Pancheva, E. Semi-selfsimilar extremal processes. *J. Math. Sci.* 2000, 99, 1306-1316.
[50] Pancheva, E; Kolkovska, E. T.; Jordanova, P. K. Random time changed extremal processes, preprint, 2003 (to appear in *Theor. Probab. Appl.* 2005).
[51] Pillai, R. N. Semistable laws as limit distributions. *Ann. Math. Statist.* 1971, 42, 780–783.
[52] Pillai, R. N. Semi-α-Laplace distributions. *Comm. Statist. Theor. Meth.* 1985, 14, 991–1000.
[53] Pillai, R. N. Semi-Pareto processes. *J. Appl. Probab.* 1991, 28, 461-465.
[54] Pillai, R. N. ; Anil, V. Symmetric stable, α-Laplace, Mittag-Leffler and related laws and processes and the integrated Cauchy functional equation. *J. Ind. Statist. Assoc.* 1996, 34, 97-103.
[55] Pillai, R. N. ; Jayakumar, K. Discrete Mittag-Leffler distributions. *Statist. Probab. Lett.* 1995, 23, 271–274.
[56] Pillai, R. N. ; Sandhya, E. Geometric sums and Pareto law in reliability theory. *IAPQR Trans.* 1996, 21, 137-142.
[57] Rachev, S. T. Rate of convergence for maxima of random arrays with applications to stock returns. *Statist. Decisions.* 1993, 11, 279–288.
[58] Rachev, S. T. ; Resnick, S. Max geometric infinite divisibility and stability. *Comm. Statist. Stoch. Models.* 1991, 7, 191–218.
[59] Ramachandran, B. On geometric stable laws, a related property of stable processes, and stable densities of exponent one. *Ann. Inst. Statist. Math.* 1997, 49, 299–313.
[60] Ramachandran, B. ; Lau, K. S. *Functional Equations in Probability Theory*, Academic Press, New York, 1991.
[61] Ramachandran, B. ; Rao, C. R. Some results on characteristic functions and characterizations of normal and generalized stable laws. *Sankhya-A.* 1968, 30, 125–140.
[62] Renyi, A. A characterization of the Poisson process. *Collected Papers of Alfred Renyi.* Vol-1, Academic Press, New York, 1956; Vol-1.
[63] Sandhya, E. *Geometric Infinite Divisibility and Applications.* Unpublished Ph.D. Thesis, University of Kerala, January 1991, 1991 (*a*).
[64] Sandhya, E. On geometric infinite divisibility, *p*-thinning and Cox processes. *J. Kerala Statist. Assoc.* 1991 (*b*), 7, 1-10.
[65] Sandhya, E. On a generalization of geometric infinite divisibility. *Proc. 8th Kerala Science Congress.* January-1996, 355–357.
[66] Sandhya, E. ; Pillai, R. N. On geometric infinite divisibility. *J. Kerala Statist. Assoc.* 1999, 10, 1-7.
[67] Sandhya, E. ; Satheesh, S. On the membership of semi-α-Laplace laws in class-L. *J. Ind. Statist. Assoc.* 1996, 34, 77–78.
[68] Satheesh, S. *Stability of Random Sums and Extremes*, Unpublished Ph.D Thesis, Cochin University of Science and Technology, July 2001, 2001.
[69] Satheesh, S. Another look at random infinite divisibility. *Statist. Meth.* 2004, 6, 123–144.
[70] Satheesh, S. Aspects of randomization in infinitely divisible and max-infinitely divisible laws. *ProbStat Models.* 2002, 1, 7–16.
[71] Satheesh, S. A supplement to the Bose-Dasgupta-Rubin (2002) review of infinitely divisible laws and processes. 2003. *http://arxiv.org/abs/math.PR/0305126.*

[72] Satheesh, S. Why there are no gaps in the support of infinitely divisible discrete laws? *ProbStat Models*. 2004 (*b*), 3, 1–7.

[73] Satheesh, S. ; Nair, N. U. Some classes of distributions on the non-negative lattice. *J. Ind. Statist. Assoc.* 2002 (*a*), 40, 41–58.

[74] Satheesh, S. ; Nair, N. U. A Note on maximum and minimum stability of certain distributions. *Calcutta Statist. Assoc. Bull.* 2002 (*b*), 53, 249-252.

[75] Satheesh, S. ; Nair, N. U. ; Sandhya, E. Stability of random sums. *Stoch. Modeling and Appl.* 2002, 5, 17–26.

[76] Satheesh, S. ; Sandhya, E. Infinite divisibility and max-infinite divisibility with random sample size. *Statist. Meth.* 2003, 5, 126–139.

[77] Satheesh, S. ; Sandhya, E. AR(1) schemes with semi-stable marginals, 2006, *http://arxiv.org/abs/math.PR/0602286*.

[78] Scher, H. ; Lax, M. Stochastic transport in a disordered solid–1. *Theor. Phys. Rev. B.* 1973, 7, 4491–4502.

[79] Shimizu, R. On the domain of partial attraction of semi-stable distributions. *Ann Inst. Statist. Math.* 1970, 22, 245–255.

[80] Sreehari, M. Maximum Stability and a generalization. *Statist. Probab. Lett.* 1995, 23, 339–342.

[81] Steutel, F. W. ; van Harn, K. Discrete analogues of self-decomposability and stability. *Ann. Probab.* 1979, 7, 893-899.

[82] Szasz, D. On classes of limit distributions for sums of a random number of identically distributed independent random variables. *Theor. Probab. Appl.* 1972, 17, 401-415.

[83] Uchaikin, V. V. ; Zolotarev, V. M. *Chance and Stability, Stable Distributions and Their Applications*, VSP, Utrecht, 1999.

[84] Voorn, W. J. Characterizations of the logistic and log-logistic distributions by extreme value related stability with random sample size. *J. Appl. Probab.* 1987, 24, 838-851.

In: Focus on Probability Theory
Editor: Louis R. Velle, pp. 73-89

ISBN 1-59454-474-3

Chapter 4

CONDITIONAL EQUI-CONCENTRATION OF TYPES*

M. Grendar†
Department of Mathematics, FPV UMB,
Tajovskeho 40, 974 01 Banska Bystrica, Slovakia,
Institute of Mathematics and CS, Slovak Academy of Sciences (SAS)
and Institute of Measurement Science of SAS

To Mar, in memoriam

Abstract

Conditional Equi-concentration of Types on I-projections is presented. It provides an extension of Conditional Weak Law of Large Numbers to the case of several I-projections. Also a multiple I-projections extension of Gibbs Conditioning Principle is developed. μ-projection variants of the probabilistic laws are stated. Implications of the results for Relative Entropy Maximization, Maximum Probability, Maximum Entropy in the Mean and Maximum Rényi-Tsallis Entropy methods are discussed.

Key Words and Phrases: multiple I-projections, Conditional Weak Law of Large Numbers, Gibbs Conditioning Principle, μ-projection, Maximum Probability method, MaxProb/MaxEnt convergence, Maximum Rényi-Tsallis Entropy method, Maximum Entropy in the Mean, triple point

2000 Mathematics Subject Classification Primary 62B15; Secondary 94A15, 62J20

1 Terminology and Notation

Let $\{X\}_{l=1}^{n}$ be a sequence of independently and identically distributed random variables with a common law (measure) on a measurable space. Let the measure be concentrated on finite number m of atoms from the set $\mathrm{X} = \{x_1, x_2, \ldots, x_m\}$ called support or alphabet. Let q_i denote the probability (measure) of i-th element of X; q will be assumed strictly positive

*This work was supported by VEGA 1/0264/03 grant. Valuable discussions with George Judge, Brian R. La Cour, Alberto Solana-Ortega, Ondrej Such and Viktor Witkovský are gratefully acknowledged. Lapses are mine.

†E-mail address: marian.grendar@savba.sk

and called source or generator. Let $P(X)$ be a set of all probability mass functions (pmf's) on X.

A type (also called n-type, empirical measure, frequency distribution or occurrence vector) induced by a sequence $\{X\}_{l=1}^{n}$ is pmf $\nu^n \in P(X)$ whose i-th element ν_i^n is defined as: $\nu_i^n = n_i/n$ where $n_i = \sum_{l=1}^{n} I(X_l = x_i)$; there $I(\cdot)$ is the indicator function. Multiplicity $\Gamma(\nu^n)$ of type ν^n is: $\Gamma(\nu^n) = n!/\prod_{i=1}^{m} n_i!$.

Let $\Pi \subseteq P(X)$. Let P_n denote a subset of $P(X)$ which consists of all n-types. Let $\Pi_n = \Pi \cap P_n$.

On $P(X)$ topology induced by the standard topology on R^m is assumed.

μ-projection $\hat{\nu}^n$ of q on $\Pi_n \neq \emptyset$ is defined as: $\hat{\nu}^n = \arg\sup_{\nu^n \in \Pi_n} \pi(\nu^n; q)$, where $\pi(\nu^n; q) = \Gamma(\nu^n) \prod (q_i)^{n\nu_i^n}$. Alternatively, the μ-projection can be defined as $\hat{\nu}^n = \arg\sup_{\nu^n \in \Pi_n} \pi(\nu^n | \nu^n \in \Pi_n; q)$, where $\pi(\nu^n|\nu^n \in \Pi_n; q)$ denotes the conditional probability that if an n-type belongs to Π_n then it is just the type ν^n. μ-projection can be also equivalently defined as the supremum of posterior probability, cf. [14].

I-projection $\hat{p}$ of q on Π is $\hat{p} = \arg\inf_{p \in \Pi} I(p||q)$, where $I(p||q) = \sum_X p_i \log \frac{p_i}{q_i}$ is the I-divergence (also known as Kullback-Leibler distance, $\pm$ relative entropy).

$\pi(\nu^n \in A | \nu^n \in B; q)$ will denote the conditional probability that if a type drawn from $q \in P(X)$ belongs to $B \subseteq \Pi$ then it belongs to $A \subseteq \Pi$.

2 Boltzmann Jaynes Inverse Problem and Conditional Law of Large Numbers

Having the terminology introduced, Boltzmann Jaynes Inverse Problem (BJIP) can be stated as follows: there is the source q and a set Π_n of n-types. It is necessary to select an n-type (one or more) from the set Π_n. To solve BJIP it is necessary to provide an algorithm for selection of type from Π_n when the information-quadruple $\{X, n, q, \Pi_n\}$ and nothing else is supplied. Clearly, if Π_n contains more than one type, BJIP becomes an under-determined and in this sense ill-posed problem.

Usually, BJIP is solved by means of Relative Entropy Maximization (REM/MaxEnt) method. This is mostly done for $n \to \infty$. In this case the set of types Π_n effectively turns into a set of probability mass functions Π.

Typically, Π is defined by moment consistency constraints (mcc) of the following form [1]: $\Pi_{mcc} = \{p : \sum_{i=1}^{m} p_i u_i = a, \sum_{i=1}^{m} p_i = 1\}$, where $a \in R$ is a given number, u is a given vector. The feasible set Π_{mcc} which mcc define is convex and closed. I-projection $\hat{p}$ of q on Π_{mcc} is unique and belongs to the exponential family of distributions; $\hat{p}_i = k(\lambda) q_i e^{-\lambda u_i}$, where $k(\lambda) = 1/\sum_{i=1}^{m} q_i e^{-\lambda u_i}$, and λ is such that $\hat{p}$ satisfies mcc.

In the case of BJIP with Π_{mcc}, or in general for any closed, convex, rare set Π, application of REM/MaxEnt method is justified by Conditional Weak Law of Large Numbers (CWLLN). CWLLN, in its textbook form, reads [4]:

CWLLN. *Let X be a finite set. Let Π be a closed, convex set which does not contain q. Let $n \to \infty$. Then for $\varepsilon > 0$ and $i = 1, 2, \ldots, m$,*

$$\lim_{n \to \infty} \pi(|\nu_i^n - \hat{p}_i| \leq \varepsilon \,|\, \nu^n \in \Pi; q) = 1.$$

[1] In the simplest case of single non-trivial constraint.

CWLLN says that if types are confined to the set Π then they asymptotically conditionally concentrate on the I-projection $\hat{p}$ of the source of types q on the set Π. Stated, informally, from another side: if a source q is confined to produce types from convex and closed Π it is asymptotically conditionally 'almost impossible' to find a type other than the one which has the highest/supremal value of relative entropy with respect to q.

Conditional Weak Law of Large Numbers emerged from a series of works which include [34], [1], [40], [28], [21], [39], [41], [38], [36], [5], [9], [8], [30], [29]. For new developments see [23].

An information-theoretic proof (see [4]) of CWLLN utilizes so-called Pythagorean theorem (cf. [2]), Pinsker inequality and standard inequalities for factorial. The Pythagorean theorem is known to hold for closed convex sets. Alternatively, CWLLN can be obtained as a consequence of Sanov's Theorem (ST). The ST-based proof of CWLLN will be recalled here. First, Sanov's Theorem and its proof (adapted from [7], [4]).

Sanov's Theorem. *Let* X *be finite. Let* $\mathrm{A} \subseteq \Pi$ *be an open set. Then*

$$\lim_{n\to\infty} \frac{1}{n} \log \pi(\nu^n \in \mathrm{A}) = -I(\hat{p}||q),$$

where $\hat{p}$ *is an I-projection of* q *on* A.

Proof. [4], [7] $\pi(\nu^n \in \mathrm{A}) = \sum_{\nu^n \in \mathrm{A}} \pi(\nu^n; q)$. Upper and lower bounds on $\pi(\nu^n; q)$ (recall proof of the Lemma at Appendix):

$$\left(\frac{m}{n}\right)^m \prod_{i=1}^{m} \left(\frac{q_i}{\nu_i^n}\right)^{n\nu_i^n} < \pi(\nu^n; q) \le \prod_{i=1}^{m} \left(\frac{q_i}{\nu_i^n}\right)^{n\nu_i^n}.$$

$\sum_{\nu^n \in \mathrm{A}} \pi(\nu^n; q) < N \prod_{i=1}^{m} (\frac{q_i}{\hat{\nu}_i^n})^{n\hat{\nu}_i^n}$, where N stands for number of all n-types and $\hat{\nu}_i^n$ is an I-projection of q on $\mathrm{A}_n = \mathrm{A} \cap \mathrm{P}_n$ (i.e., any of the n-types which attain supremal value of $\prod_{i=1}^{m} (\frac{q_i}{\hat{\nu}_i^n})^{n\hat{\nu}_i^n}$). N is smaller than $(n+1)^m$.

Thus

$$\frac{1}{n}\left(n \sum_{i=1}^{m} \hat{\nu}_i^n \log \frac{q_i}{\hat{\nu}_i^n} + m(\log m - \log n)\right) < \frac{1}{n} \log \pi(\nu^n \in \mathrm{A})$$

$$< \frac{1}{n}\left(n \sum_{i=1}^{m} \hat{\nu}_i^n \log \frac{q_i}{\hat{\nu}_i^n} + m \log(n+1)\right).$$

Since A is by the assumption open and under the maintained assumption of strictly positive q it is also continuous, $\lim_{n\to\infty} \sum \hat{\nu}_i^n \log \frac{q_i}{\hat{\nu}_i^n} = \sum \hat{p}_i \log \frac{q_i}{\hat{p}_i}$, where $\hat{p}$ is an I-projection of q on A. Thus, for $n \to \infty$ the upper and lower bounds on $\frac{1}{n} \log \pi(\nu^n \in \mathrm{A})$ collapse into $\sum_{i=1}^{m} \hat{p}_i \log \frac{q_i}{\hat{p}_i}$. □

A proof of CWLLN. [7] Let $\mathrm{A} = \{p : |p_i - \hat{p}_i| > \varepsilon, i = 1, 2, \dots, m\}$. Then ST can be applied to it, leading $\lim_{n\to\infty} \frac{1}{n} \log \pi(\nu^n \in \mathrm{A} | \nu^n \in \Pi; q) = -(I(\hat{p}_{\mathrm{A}}||q) - I(\hat{p}_{\Pi}||q))$. Since $I(\hat{p}_{\mathrm{A}}||q) - I(\hat{p}_{\Pi}||q) > 0$ and since the set Π admits unique I-projection (the uniqueness arises from the fact that the set is convex and closed, and $I(\cdot||\cdot)$ is convex), the proof is complete. □

CWLLN can be viewed as a special case of a stronger result, which is commonly known as Gibbs Conditioning Principle (GCP), see Sect. 5.

3 Motivation and Programme

Frequency moment constraints considered by physicists (see for instance [33]) define a non-convex feasible set of probability distributions which in general can admit multiple I-projections. This work builds upon [12], [27], [13], [15], [16] and aims to develop an extension of CWLLN and Gibbs Conditioning Principle to the case of multiple I-projections.

It has also another goal: to introduce concept of μ-projection and to formulate μ-projection variants of the probabilistic laws. They, among other things, allow for a more elementary reading than their I-projection counterparts. At the same time they provide a probabilistic justification of Maximum Probability method [11].

The paper is organized as follows: in the next section some basic questions regarding asymptotic behavior of conditional probability are posed. Two illustrative examples are then used to introduce Conditional Equi-concentration of Types on I-projections. Next, an extension of Gibbs Conditioning Principle - the stronger form of CWLLN - is provided. Asymptotic identity of I-projections and μ-projections is discussed in Section 6 and μ-variants of the probabilistic laws are presented afterwards. Implications of the results for Maximum Entropy, Maximum Probability and Maximum Rényi-Tsallis Entropy methods are drawn at Section 8. Section 9 mentions in passing other related results: r-tuple extension of CWLLN and Bayesian Conditional Law of Large Numbers. Section 10 summarizes the paper. Appendix contains a sketch of proof of ICET and of Extended GCP. It is also shown there that concentration of types can in some sense happen also on isolated I-projections, provided that they are rational.

4 Conditional Equi-concentration of Types

What happens when Π admits multiple I-projections? Do the conditional concentration of types happens on them? If yes, do the type concentrate on each of them? If yes, what is the proportion? In order to address these questions, it is instrumental to consider a couple of examples.

Example 1. [13] Let $\Pi = \{p : \sum_{i=1}^{m} p_i^{\alpha} - a = 0, \sum_{i=1}^{m} p_i - 1 = 0\}$, where $\alpha, a \in R$. Note that the first constraint, known as frequency constraint, is non-linear in p and Π is for $|\alpha| > 1$ non-convex.

Let $\alpha = 2$, $m = 3$ and $a = 0.42$ (the value was obtained for $p = [0.5\ 0.4\ 0.1]$). Then there are the following three I-projections of uniform distribution $q = [1/3\ 1/3\ 1/3]$ on Π: $\hat{p}_1 = [0.5737\ 0.2131\ 0.2131]$, $\hat{p}_2 = [0.2131\ 0.5737\ \ 0.2131]$ and $\hat{p}_3 = [0.2131\ 0.2131\ 0.5737]$ (see [15]). Note that they form a group of permutations. As it will become clear later, it suffices to investigate convergence to say $\hat{p}_1$.

For $n = 30$ there are only two groups of types in Π: G1 comprises $[0.5666\ 0.2666\ 0.1666]$ and five other permutations; G2 consists of $[0.5\ 0.4\ 0.1]$ and the other five permutations. So, together there are 12 types.

Value of the square of the Euclidean distance δ between ν and $\hat{p}_1$ attains its minimum $\delta_{G1} = 0.0051$ within G1 group for the following two types: $[0.5666\ 0.2666\ \ 0.1666]$, $[0.5666\ 0.1666\ 0.2666]$. Within G2 the smallest $\delta_{G2} = 0.0532$ is attained by $[0.5\ 0.4\ 0.1]$ and $[0.5\ 0.1\ 0.4]$.

Thus, if $\varepsilon = \varepsilon_1$ is chosen so that the ε-ball $B(\hat{p}_1, \varepsilon_1)$ centered at $\hat{p}_1$ contains only the two types from G1 (which at the same time guarantees that $\hat{p}_1$ is the only I-projection in the ball), then $\pi(\nu \in B(\hat{p}_1, \varepsilon_1) | \nu \in \Pi) = 2 \cdot 0.1152 = 0.2304$. In words: probability that if q generated a type from Π than the type falls into the ball containing only types which are closest to the I-projection is 0.2304. If $\varepsilon = \varepsilon_2$ is chosen so that also the two types from G2 are included in the ball and also so that it is the only I-projection in the ball (any $\varepsilon_2 \in (\sqrt{0.0532}, \sqrt{0.1253})$ satisfies both the requirements), then $\pi(\nu^n \in B(\hat{p}_1, \varepsilon_2) | \nu^n \in \Pi) = \frac{1}{3}$.

For $n = 330$ there are four groups of types in Π: G1, G2 and a couple of new one: G3 consists of $[0.4727\ 0.4333\ 0.0939]$ and all its permutations; G4 comprises the type $[0.5727\ 0.2333\ 0.1939]$ and its permutations. Hence, the total number of types from Π which are supported by random sequences of size $n = 330$ is 24.

δ_{G3} for the two types from G3 which are closest to $\hat{p}_1$ is 0.0729. The smallest $\delta_{G4} = 0.00077$ is attained by $[0.5727\ 0.2333\ 0.1939]$ and by $[0.5727\ 0.1939\ 0.2333]$. Thus, clearly, the two types from G4 have the smallest Euclidean distance to $\hat{p}_1$ among all types from Π which are based on samples of size $n = 330$. Again, setting ε such that the ball $B(\hat{p}_1, \varepsilon)$ contains only the two types which are closest to $\hat{p}_1$ leads to the 0.261 value of the conditional probability. Note the important fact, that the probability has risen, as compared to the corresponding value 0.2304 for $n = 30$.

Moreover, if ε is set such that besides the two types from G4 also the second closest types (i.e. the two types from G1) are included in the ball then the conditional probability is indistinguishable from $\frac{1}{3}$. Hence, there is virtually no conditional chance of observing any of the remaining 4 types. The same holds for the types which concentrate around $\hat{p}_2$ or $\hat{p}_3$. Thus, in total, a half of the 24 types is almost impossible to observe.

The Example illustrates, that the conditional probability of finding a type which is close (in the Euclidean distance) to one of the three I-projections goes to $\frac{1}{3}$.

Example 2. [13] Let $\Pi = \Pi_1 \cup \Pi_2$, where $\Pi_j = \{p : \sum_{i=1}^{m} p_i x_i = a_j;\ \sum_{i=1}^{m} p_i = 1\}, j = 1, 2$. Thus Π is union of two sets, each of whose is given by the moment consistency constraint. If q is chosen to be the uniform distribution, then values a_1, a_2 such that there will be two different I-projections of the uniform q on Π with the same value of I-divergence (as well as of the Shannon's entropy) can be easily found. Indeed, for any $a_1 = \mu + \Delta$, $a_2 = \mu - \Delta$, where $\mu = EX$ and $\Delta \in (0, (X_{\max} - X_{\min})/2)$, $\hat{p}_1$ is just a permutation of $\hat{p}_2$, and as such attains the same value of Shannon's entropy. To see that types which are based on random samples of size n from Π indeed concentrate on the I-projections with equal measure note, that for any n to each type in Π_1 corresponds a unique permutation of the type in Π_2. Thus, types in ε-ball with center at $\hat{p}_1$ have the same conditional probabilities π as types in the ε-ball centered at $\hat{p}_2$. This, together with convexity and closed-ness of both Π_j, for which the conditional concentration of types on the respective I-projection is established by CWLLN, directly implies that

$$\lim_{n \to \infty} \pi(\nu \in B(\hat{p}_j, \varepsilon) | \nu^n \in \Pi) = \frac{1}{2} \qquad j = 1, 2.$$

Conditional Equi-concentration of Types on I-projections (ICET) attempts to capture behavior of the conditional measure which the above Examples illustrate. To this end, notion of the proper I-projection will be needed.

I-projection $\hat{p}$ of q on Π will be called proper if $\hat{p}$ is not isolated point of Π.

***I*CET.** *Let* X *be finite. Let there be* k *proper I-projections* $\hat{p}^1, \hat{p}^2, \ldots, \hat{p}^{\mathrm{k}}$ *of* q *on* Π. *Let* $\varepsilon > 0$ *be such that for* $j = 1, 2, \ldots, \mathrm{k}$ $\hat{p}^j$ *is the only proper I-projection of* q *on* Π *in the ball* $B(\hat{p}^j, \varepsilon)$. *Let* $n \to \infty$. *Then for* $j = 1, 2, \ldots, \mathrm{k}$,

$$\pi(\nu^n \in B(\varepsilon, \hat{p}^j) | \nu^n \in \Pi; q) = 1/\mathrm{k}.$$

ICET says, informally, that source/generator q, when confined to produce types from a set Π, - as n gets large - hides itself behind any of the proper I-projections equally likely.

Expressed in Statistical Physics terminology ICET says that each of equilibrium points (I-projections) is asymptotically conditionally equally possible. The Conditional Equi-concentration of Types 'phenomenon' resembles the triple point phenomenon of Thermodynamics.

5 Gibbs Conditioning Principle and its Extension

Gibbs conditioning principle (cf. [5], [8], [29]) - also known as the stronger form of CWLLN - complements CWLLN by stating that:

GCP. *Let* X *be a finite set. Let* Π *be closed, convex set. Let* $n \to \infty$. *Then for a fixed* t,

$$\lim_{n \to \infty} \pi(X_1 = x_1, \ldots, X_t = x_t | \nu^n \in \Pi; q) = \prod_{l=1}^{t} \hat{p}_{x_l}.$$

GCP, says, very informally, that if the source q is confined to produce sequences which lead to types in a set Π then elements of any such sequence (of fixed length t) behave asymptotically conditionally as if they were drawn identically and independently from the I-projection of q on Π - provided that the last is unique (among other things).

GCP was developed at [5] under the name of conditional quasi-independence of outcomes. Later on, it was brought into more abstract form in large deviations literature, where it also obtained the GCP name (cf. [8], [29]). A simple proof of GCP can be found at [7]. GCP is proven also for continuous alphabet (cf. [21], [7], [8]).

The following theorem provides an extension of GCP to the case of multiple I-projections.

EGCP. *Let there be* k *proper I-projections* $\hat{p}^1, \hat{p}^2, \ldots, \hat{p}^{\mathrm{k}}$ *of* q *on* Π. *Then for a fixed* t *and* $n \to \infty$,

$$\pi(X_1 = x_1, \ldots, X_t = x_t | \nu^n \in \Pi; q) = \frac{1}{\mathrm{k}} \sum_{j=1}^{\mathrm{k}} \prod_{l=1}^{t} \hat{p}^j_{x_l}.$$

For $t = 1$ Extended Gibbs Conditioning Principle (EGCP) says that the conditional probability of a letter is asymptotically given by the equal-weight mixture of proper I-projection probabilities of the letter. For a general sequence, EGCP states that the conditional probability of a sequence is asymptotically equal to the mixture of joint probability distributions. Any (j-th) of the k joint distributions is such as if the sequence was iid distributed according to a (j-th) proper I-projection.

6 Asymptotic Identity of μ-Projections and I-Projections

At ([11], Thm 1 and its Corollary, aka MaxProb/MaxEnt Thm) it was shown that maximum probability type converges to I-projection; provided that Π is defined by a differentiable constraints. A more general result which states asymptotic identity of μ-projections and I-projections for general set Π was presented at [16].

MaxProb/MaxEnt. *Let* X *be finite set. Let* M_n *be set of all μ-projections of q on Π_n. Let* I *be set of all I-projections of q on Π. For $n \to \infty$,* $\mathrm{M}_n = \mathrm{I}$.

Since $\pi(\nu^n; q)$ is defined for $\nu^n \in \mathrm{Q}^m$, μ-projection can be defined only for Π_n when n is finite. The Thm permits to define a μ-projection $\hat{\nu}$ also on Π: $\hat{\nu} = \arg\sup_{r\in\Pi} - \sum_{i=1}^m r_i \log \frac{r_i}{q_i}$. The μ-projections of q on Π and I-projections of q on the same set Π are undistinguishable.

It is worth highlighting that for a finite n, μ-projections and I-projections of q on Π_n are in general different. This explains why μ-form of the probabilistic laws deserves to be stated separately of the I-form; though formally they are undistinguishable. Thus, the MaxProb/MaxEnt Thm (in its new and to a smaller extent also in its old version) permits directly to state μ-projection variants of CWLLN, GCP, ICET and EGCP: μCWLLN, μGCP, μCET and Boltzmann Conditioning Principle (BCP).

7 μ-Variants of the Probabilistic Laws

μ-variant of CWLLN reads:

μCWLLN. *Let* X *be a finite set. Let Π be closed, convex set. Let $n \to \infty$. Then for $\varepsilon > 0$ and $i = 1, 2, \ldots, m$,*

$$\lim_{n\to\infty} \pi(|\nu_i^n - \hat{\nu}_i| < \varepsilon | \nu^n \in \Pi; q) = 1.$$

Core of μCWLLN can be loosely expressed as: *types, when confined to a set Π, conditionally concentrate on the asymptotically most probable type $\hat{\nu}$.*

μ-projection $\hat{\nu}$ of q on Π will be called proper if $\hat{\nu}$ is not isolated point of Π.

μCET. *Let* X *be finite. Let there be* k *proper μ-projections $\hat{\nu}^1, \hat{\nu}^2, \ldots, \hat{\nu}^{\mathrm{k}}$ of q on Π. Let $\varepsilon > 0$ be such that for $j = 1, 2, \ldots, \mathrm{k}$ $\hat{\nu}^j$ is the only proper μ-projection of q on Π in the ball $B(\hat{\nu}^j, \varepsilon)$. Let $n \to \infty$. Then for $j = 1, 2, \ldots, \mathrm{k}$,*

$$\pi(\nu^n \in B(\varepsilon, \hat{\nu}^j) | \nu^n \in \Pi; q) = 1/\mathrm{k}.$$

Core of μ-variant of the Conditional Equi-concentration of Types states, loosely, that types conditionally concentrate on each of the asymptotically most probable types in equal measure.

μGCP. *Let* X *be a finite set. Let Π be closed, convex set. Let $n \to \infty$. Then for a fixed t,*

$$\lim_{n\to\infty} \pi(X_1 = x_1, \ldots, X_t = x_t | \nu^n \in \Pi; q) = \prod_{l=1}^{t} \hat{\nu}_{x_l}.$$

μ-variant of EGCP deserves a special name. It will be called Boltzmann Conditioning Principle (BCP).

BCP. *Let there be* k *proper* μ*-projections* $\hat{\nu}^1, \hat{\nu}^2, \ldots, \hat{\nu}^k$ *of* q *on* Π*. Then for a fixed* t *and* $n \longrightarrow \infty$*,*

$$\pi(X_1 = x_1, \ldots, X_t = x_t | \nu^n \in \Pi; q) = \frac{1}{k} \sum_{j=1}^{k} \prod_{l=1}^{t} \hat{\nu}^j_{x_l}.$$

8 Implications

The results have some bearing for application of REM, MaxProb and Maximum Rényi-Tsallis Entropy methods to Boltzmann Jaynes Inverse Problem.

8.1 I- or μ-projection? MaxEnt or MaxProb?

With μ-projection Maximum Probability method (MaxProb, [11]) is associated. Given the BJIP information-quadruple $\{X, n, q, \Pi_n\}$, MaxProb prescribes to select from Π_n type(s) which has the supremal/maximal probability $\pi(\nu^n; q)$[2].

μ-projections and I-projections are asymptotically indistinguishable. In plain words: for $n \longrightarrow \infty$ the Relative Entropy Maximization method (REM/MaxEnt) (either in its Jaynes' [24], [25] or Csiszár's interpretation [6]) selects the same distribution(s) as MaxProb (in its more general form which instead of the maximum probable types selects supremum-probable μ-projections). This result (in the older form, [11]) was at [11] *interpreted* as saying that REM/MaxEnt can be viewed as an asymptotic instance of the simple and self-evident Maximum Probability method.

Alternatively, [35] suggests to view REM/MaxEnt as a separate method and hence to read the MaxProb/MaxEnt Thm as claiming that REM/MaxEnt asymptotically coincides with MaxProb. If one adopts this interesting and legitimate view then it is necessary to face the fact that if n is finite, the two methods in general differ. This would open new questions. Among them also: MaxEnt/REM or MaxProb? (i.e., I- or μ-projection?) This is too delicate a question to be answered by one sentence. Let us note, only, that unless $n \longrightarrow \infty$ entropy ignores multiplicity.

8.2 I/μ- or τ-projection? MaxEnt/MaxProb or MaxTent?

The previous question (i.e., MaxEnt or MaxProb?) is a problem of drawing an interpretational consequences from two variants of the same probabilistic laws, and in this sense it can be viewed as an 'internal problem' of MaxEnt and MaxProb. From outside, from the point of view of the Maximum Rényi-Tsallis Entropy method (maxTent, [37], [26], [3]) MaxProb and MaxEnt can be viewed as 'twins'.

maxTent is to the best of our knowledge intended by its proponents for selection of probability distribution(s) under the setting of BJIP with Π defined by X-frequency moment constraints (cf. [15]). It is not known whether such a feasible set Π admits unique distribution with maximal value of Rényi-Tsallis entropy (called τ-projection at [15]) as it is also

[2] A technique for determination of μ-projections was suggested at [17].

not known whether I-projection on such a set is unique or not. The non-uniqueness makes it difficult to relay upon CWLLN when one wants to draw from an established non-identity of τ and I-projection conclusion that maxTent method violates CWLLN, cf. [27]. At [15] this difficulty has been avoided by considering an instance of the X-frequency constraints where the feasible set reduced into a convex set. Since τ- and I-projection on the set were shown to be different, CWLLN directly implies that maxTent in this case selects asymptotically conditionally improbable distribution. The Example below (taken from [15]) illustrates the point.

Example 3. [15] Let $\Pi = \{p : \sum_{i=1}^{3} p_i^2(x_i - b) = 0, \sum_{i=1}^{3} p_i - 1 = 0\}$. Let $\mathrm{X} = [-2\ \ 0\ \ 1]$ and let $b = 0$. Then $\Pi = \{p : p_3^2 = 2p_1^2, \sum p_i - 1 = 0\}$ which effectively reduces to $\Pi = \{p : p_2 = 1 - p_1(1+\sqrt{2}), p_3 = \sqrt{2}p_1\}$. The source q is assumed to be uniform u.

The feasible set Π is convex. Thus I-projection $\hat{p}$ of u on Π is unique, and can be found by direct analytic maximization to be $\hat{p} = [0.2748\ \ 0.3366\ \ 0.3886]$. Straightforward maximization of Rényi-Tsallis' entropy leads to maxTent pmf $\hat{p}_T = [0.2735\ \ 0.3398\ \ 0.3867]$, which is different than $\hat{p}$.

Convexity of the feasible set guarantees uniqueness of the I-projection, and consequently allows to invoke CWLLN to claim that any pmf from Π other than the I-projection has asymptotically zero conditional probability that it will be generated by u.

Obviously, ICET permits to show the fatal flow of maxTent in a more direct and more general way.

9 Further Results

Some further results related to asymptotic concentration of conditional probability are contained in this Section.

9.1 r-tuple ICET/CWLLN and MEM/GME Methods

Maximum Entropy in the Mean method (MEM), or its discrete-case relative, Generalized Maximum Entropy (GME) method, are interesting extensions of the standard REM/MaxEnt method[3]. Though, usually a hierarchical structure of the methods is highlighted, here a different feature of the method(s) will appear to be important.

First, Golan-Judge-Miller ill-posed inverse problem (GJMIP) has to be introduced. Its simple instance can be described as follows: Let there be two independent sources q^1, q^2 of sequences and hence types. Let X, Y be support of the first, second source, respectively. Let a set C_n comprise *pairs* of the types $[\nu^{n,1}, \nu^{n,2}]$ which were drawn *at the same time*. GJMIP amounts to selection of specific pair(s) of types from C_n when the information $\{\mathrm{X}, \mathrm{Y}, n, q^1 \perp q^2, \mathrm{C}_n\}$ is supplied.

Example 4. An example of GJMIP. Let $\mathrm{X} = \mathrm{Y} = [1\ 2\ 3]$. Let $q^1 = q^2 = [1/3\ 1/3\ 1/3]$; $q^1 \perp q^2$; $(q^1 \mapsto \nu^{n,1}) \wedge (q^2 \mapsto \nu^{n,2})$. Let $n = 100$, $\mathrm{C}_n = \{[\nu^{n,1}, \nu^{n,2}] : \sum_{i=1}^{3} \nu_i^{n,1} x_i + \nu_i^{n,2} y_i = 4; \sum_{i=1}^{3} \nu_i^{n,1} = 1; \sum_{i=1}^{3} \nu_i^{n,2} = 1\}$. Given this information, it is necessary to select a pair (one or more) of types from C_n.

[3]For a tutorial on MEM see [22]. GME was introduced at [10], see also [31].

Since throughout the paper discrete and finite alphabet is assumed, GME will be considered instead of MEM, in what follows. The important feature of GME is that it selects jointly and independently drawn pairs (or r-tuples) of types/pmfs. Thus, it is suitable for application at the GJMIP context. An r-tuple extension of CWLLN (rCWLLN) provides a probabilistic justification to the GME, at the GJMIP context.

Given GJMIP information, GME selects from the feasible set of the pairs of pmfs the one $[\hat{p}^1, \hat{p}^2]$ (or more) which maximizes sum of the relative entropies with respect to q^1, q^2; respectively.

($r = 2$)-tuple CWLLN. *Assume a GJMIP. Let C be convex, closed set. Let $B([\hat{p}^1, \hat{p}^2], \varepsilon)$ be an ε-ball centered at the pair*

$$[\hat{p}^1, \hat{p}^2] = \arg \sup_{[p^1, p^2] \in C} \sum_{i=1}^{m^1} p_i^1 \log \frac{p_i^1}{q_i^1} + \sum_{j=1}^{m^2} p_j^2 \log \frac{p_j^2}{q_j^2}.$$

Let $n \to \infty$. Then

$$\pi([\nu^{n,1}, \nu^{n,2}] \in B([\hat{p}^1, \hat{p}^2], \varepsilon) \,|\, [\nu^{n,1}, \nu^{n,2}] \in C; (q^1 \mapsto \nu^{n,1}) \wedge (q^2 \mapsto \nu^{n,2}); q^1 \perp q^2) = 1$$

Proof of rCWLLN can be constructed along the same lines as the proof of CWLLN; the assumption that the pairs of sequnces/types are drawn at the same time and from independent sources is crucial for establishing the result. Similarly, r-generalization of ICET can be formulated and proven; and obviously μ-variants of the results hold true.

rCWLLN permits to rise the same objections to application of Rényi entropy based variant of GME in the GJMIP context, as those that were risen to maxTent in the BJIP context.

Needless to say, μ-variant of rCWLLN provides a probabilistic justification to MaxProb variant of GME.

9.2 Bayesian Conditional Law of Large Numbers

It is worth a brief mentioning, that there is an inverse problem which is in a sense antipodal to Boltzmann Jaynes Inverse Problem. Let us call it the β-problem, after [19].

One form of the β-problem can be formulated as follows: let there be a set Q of sources over which a prior distribution $\pi(\cdot)$ is specified. Let ν^n be an n-type drawn from a source r, not necessarily in Q. It is necessary to select a source $q \in$ Q, given the information-pentad $\{n, X, \nu^n, Q, \pi(\cdot)\}$.

Conditional Law of Large Number for Sources [18], [20] is concerned about the asymptotic behavior of posterior probability $\pi(q \in B | (q \in Q) \wedge \nu^n)$. It states that, under certain conditions, the posterior probability asymptotically piles up on the L-projection of r on Q. Hence, the particular β-problem has to be solved by L-divergence maximization method.

An application of Conditional Limit Theorem for Sources to criterion choice problem associated with the Empirical Estimation [31], [32] as well as further discussion, can be found at [19].

10 Summary

Conditional Equi-concentration of Types on I-projections – an extension of CWLLN to the case of non-unique I-projection – was presented. ICET states that the conditional concentration of types happens on each of the proper I-projections in equal measure. Also, Gibbs Conditioning Principle was enhanced to capture multiple I-projections. Extended GCP says (when $t = 1$) that conditional probability of a letter is asymptotically given by the equal-weight mixture of proper I-projection probabilities of the letter. The conditional equi-concentration/equi-probability 'phenomenon' is in our view an interesting feature of 'randomness'. It might be of some interest also for Statistical Mechanics as it resembles phase coexistence of Thermodynamics (eg. triple point of water, vapor and ice).

A general form of MaxProb/MaxEnt Thm, which states asymptotic identity of I- and μ-projections, was recalled. It permits to formulate μ-projection variants of the corresponding I-projection laws: CWLLN/GCP/ICET/EGCP. In our view, the μ-variants allow for a deeper reading than their I-projection counterparts – since the μ-laws express the asymptotic conditional behavior of types in terms of the most probable types. For instance, μ-projection variant of CWLLN says that types conditionally concentrate on the asymptotically most probable one. This is, in our view, more obvious statement than that made by I-variant of CWLLN. MaxProb/MaxEnt Theorem is also instrumental for establishing of ICET.

The main results – Conditional Equi-concentration of Types (CET) in both its I- and μ-projection form as well as Extended Gibbs Conditioning and Boltzmann Conditioning – were supplemented also by further considerations. They are summarized below.

Though μ-projections and I-projections asymptotically coincide, for a finite n they are, in general, different. In light of this fact, the asymptotic identity of μ- and I-projections can be viewed in two ways. Either as saying that 1) I-projection of q on Π is the asymptotic form of μ-projection of q on Π_n or that 2) μ-projections on Π_n and I-projections on Π_n asymptotically coincide. Regardless of the preferred view, the μ-variants of the laws provide a probabilistic justification of Maximum Probability method (MaxProb, cf. [11]) (at least in the area of Boltzmann-Jaynes inverse problem). If the second view is adopted, then, when n is finite, it is necessary to face the challenge of selecting between REM/MaxEnt method and MaxProb method.

The results have a relevance also for Maximum Rényi-Tsallis Entropy method (maxTent), which is over the last years in vogue in Statistical Physics. maxTent is to the best of our knowledge proposed as a method for solving BJIP, albeit with the feasible set Π defined by non-linear moment constraints. Since, in general, maxTent distributions (τ-projections on Π) are different than I/μ-projections on Π, ICET implies that the maxTent method selects asymptotically conditionally improbable/impossible distributions.

A straightforward extension of CWLLN/CET for r-tuples of types was also mentioned. It was noted that the extension provides a justification to the Generalized Maximum Entropy method in the area of Golan-Judge-Miller Inverse Problem.

Conditional Law of Large Numbers for Sources and its implications for the β-problem were also mentioned, in passing.

11 Appendix

11.1 MaxProb/MaxEnt

MaxProb/MaxEnt. *Let* X *be finite set. Let* M_n *be set of all* μ*-projections of* q *on* Π_n*. Let* I *be set of all I-projections of* q *on* Π*. For* $n \to \infty$, $\mathrm{M}_n = \mathrm{I}$.

Proof. [16] Necessary and sufficient conditions for $\hat{\nu}^n$ to be a μ-projection of q on Π_n are: *a)* $\pi(\hat{\nu}^n; q) \geq \pi(\nu^n; q)$, $\forall \nu^n \in \Pi_n$; *b)* whenever $\tilde{\nu}^n$ has the property *a)* then $\pi(\hat{\nu}^n; q) \leq \pi(\tilde{\nu}^n; q)$. Requirement *a)* can be equivalently stated as:

$$\left(\prod \frac{n_i!}{\hat{n}_i!}\right)^{1/n} \geq \left(\prod q_i^{n_i - \hat{n}_i}\right)^{1/n} \tag{1}$$

and *b)* similarly. Standard inequality $(n/e)^n < n! < n(n/e)^n$ (valid for $n > 6$) allows to bind the LHS of (1):

$$\frac{\prod (\nu_i^n)^{\nu_i^n}}{n^{m/n} \prod (\hat{\nu}_i^n)^{\hat{\nu}_i^n} (\prod \hat{\nu}_i^n)^{1/n}} < \text{LHS} < \frac{n^{m/n} \prod (\nu_i^n)^{\nu_i^n} (\prod \nu_i^n)^{1/n}}{\prod (\hat{\nu}_i^n)^{\hat{\nu}_i^n}} \tag{2}$$

and similar bounds can be stated in the case of the requirement *b)*[4]. Since m is by assumption finite, as $n \to \infty$ the lower and upper bounds at (2) collapse into the ratio $\prod (\nu_i^n)^{(\nu_i^n)} / \prod (\hat{\nu}_i^n)^{(\hat{\nu}_i^n)}$. Consequently, the necessary and sufficient conditions *a)*, *b)* for μ-projection turn as $n \to \infty$ into (expressed in an equivalent log-form): *i)* $\sum(\nu_i^n \log \nu_i^n - \hat{\nu}_i^n \log \hat{\nu}_i^n) \geq \sum(\nu_i^n - \hat{\nu}_i^n) \log q_i$ for all $\nu^n \in \Pi_n$; and *ii)* whenever $\tilde{\nu}^n$ has the property *i)* then $\sum \hat{\nu}_i^n \log \hat{\nu}_i^n - \tilde{\nu}_i^n \log \tilde{\nu}_i^n \geq \sum(\hat{\nu}_i^n - \tilde{\nu}_i^n) \log q_i$.

Necessary and sufficient conditions for $\hat{p}$ to be an I-projection of q on Π are the following: *I)* $\sum(p_i \log p_i - \hat{p}_i \log \hat{p}_i) \geq \sum(p_i - \hat{p}_i) \log q_i$ for all $p \in \Pi$; and *II)* whenever $\tilde{p}$ has the property *I)* then $\sum \hat{p}_i \log \hat{p}_i - \tilde{p}_i \log \tilde{p}_i \geq \sum(\hat{p}_i - \tilde{p}_i) \log q_i$.

Comparison of *i)*, *ii)* and *I)*, *II)* then completes the proof. □

11.2 *I*CET

The conditional equi-concentration of types can be seen as a consequence of Sanov's Theorem and MaxProb/MaxEnt Theorem. Indeed, Sanov's Theorem implies that the probability $\pi(\nu^n \in C; q)$ decays to zero for any open set C which excludes all of the I-projections. The asymptotic identity of I- and μ-projections shows that for $n \to \infty$, the I-projections have the same value of the probability $\pi(\nu^n; q)$.

The following is a rough attempt to make the argument a bit more formal. It relays upon MaxProb/MaxEnt Thm and the Lemma, which states a standard inequality for ratio of probabilities:

Lemma. *Let* ν^n, $\mathbf{v}^n$ *be two types from* Π_n*. Then*

$$\frac{\pi(\nu^n; q)}{\pi(\mathbf{v}^n; q)} < \left(\frac{n}{m}\right)^m \prod_{i=1}^m \frac{\left(\frac{q_i}{\nu_i^n}\right)^{n\nu_i^n}}{\left(\frac{q_i}{\mathbf{v}_i^n}\right)^{n\mathbf{v}_i^n}}$$

[4]Note that if an i-th component ν_i^n of a type is zero then it can be effectively omitted from calculations of $\pi(\nu^n; q)$. Thus, it is assumed that product operations at (1), (2) are performed on non-zero components only.

Proof. $\pi(\nu^n;q) \le \prod_{i=1}^m (\frac{q_i}{\nu_i^n})^{n\nu_i^n}$. Since for $n > 6$, $(n/e)^n < n! < n(n/e)^n$, it follows that $\pi(\nu^n;q) > \frac{1}{n_1\dots n_m}\prod_{i=1}^m (\frac{q_i}{\nu_i^n})^{n\nu_i^n}$. $n_1\dots n_m < (\frac{n}{m})^m$. □

ICET. *Let* X *be finite. Let there be* k *proper I-projections* $\hat{p}^1, \hat{p}^2, \dots, \hat{p}^k$ *of* q *on* Π. *Let* $\varepsilon > 0$ *be such that for* $j = 1, 2, \dots, k$ $\hat{p}^j$ *is the only proper I-projection of* q *on* Π *in the ball* $B(\hat{p}^j, \varepsilon)$. *Let* $n \to \infty$. *Then for* $j = 1, 2, \dots, k$,

$$\pi(\nu^n \in B(\varepsilon, \hat{p}^j) | \nu^n \in \Pi; q) = 1/k.$$

Proof. Clearly,

$$\pi(\nu^n \in B(\varepsilon, \hat{p}^j) | \nu^n \in \Pi; q \mapsto \nu^n) \le \frac{\sum_{\nu^n \in B} \pi(\nu^n;q)}{\sum_{\nu^n \in \Pi} \pi(\nu^n;q)} \tag{3}$$

$B_n(\varepsilon, \hat{p}^j) \triangleq B(\varepsilon, \hat{p}^j) \cap \Pi_n$.

Without loss of generality, let there be unique I-projection $\hat{p}^n_B$ of q on the ball $B_n(\hat{p}^j, \varepsilon)$. (Sequence of the I-projections on Π_n converges to a proper I-projection of q on Π. To an I-projection on Π which is not proper, no sequence of I_n-projections converges.) Also, without loss of generality let there be k I-projections $\hat{p}^j_{\Pi_n}$, $j = 1, 2, \dots, k$ of q on Π_n.

Let A $\triangleq B_n \backslash \{\hat{p}^n_B\}$, B $\triangleq \Pi_n \backslash \{\hat{p}^j_{\Pi_n}\}, j \neq 1$, C $\triangleq \Pi_n \backslash$B.

Then the Right-Hand Side of (3) can be rewritten as:

$$\frac{\pi(\hat{p}^n_B)}{\pi(\hat{p}^1_{\Pi_n})} \frac{1 + \frac{\sum_{\nu^n \in A} \pi(\nu^n)}{\pi(\hat{p}^n_B)}}{1 + \frac{\sum_{\nu^n \in B} \pi(\nu^n)}{\pi(\hat{p}^1_{\Pi_n})} + \frac{\sum_{\nu^n \in C} \pi(\nu^n)}{\pi(\hat{p}^1_{\Pi_n})}} \tag{4}$$

By MaxProb/MaxEnt Thm I-projections have for $n \to \infty$ the same and supremal value of $\pi(\cdot)$. This implies that $\pi(\hat{p}^n_B)/\pi(\hat{p}^1_{\Pi_n})$ converges to 1 (the case of $0/0$ limit is excluded by the supremity of $\pi(\cdot)$). The same argument implies that the first ratio in the denominator converges to $k - 1$. The Lemma implies that the ratio in the nominator as well as the second ratio in the denominator converge to zero. □

11.3 Extended GCP

EGCP. *Let* X *be a finite set. Let* Π *be such that it admits* k *proper I-projections* $\hat{p}^1, \hat{p}^2, \dots, \hat{p}^k$ *of* q *on* Π. *Then for a fixed* t,

$$\lim_{n \to \infty} \pi(X_1 = x_1, \dots, X_t = x_t | \nu^n \in \Pi; q \mapsto \nu^n) = 1/k \sum_{j=1}^{k} \prod_{l=1}^{t} \hat{p}^j_{x_l}.$$

Proof. Clearly,

$$\pi(X_1 = x_1, \dots, X_t = x_t | \nu^n \in \Pi; q \mapsto \nu^n) = \frac{\sum_{\nu^n \in \Pi} \pi(X_1 = x_1, \dots, X_t = x_t, \nu^n)}{\sum_{\nu^n \in \Pi} \pi(\nu^n;q)} \tag{5}$$

Let, in addition to partitioning used in proof of *I*CET, D $\triangleq \cup_{j=1}^k \{\hat{p}^j_{\Pi_n}\}$.

Then the RHS of (5) can be rewritten as:

$$\frac{\sum_{\nu^n \in \mathrm{D}} \pi(X_1 = x_1, \ldots, X_t = x_t, \nu^n) + \sum_{\nu^n \in \Pi_n \setminus \mathrm{D}} \pi(X_1 = x_1, \ldots, X_t = x_t, \nu^n)}{\pi(\hat{p}^1_{\Pi_n})(1 + \frac{\sum_{\nu^n \in \mathrm{B}} \pi(\nu^n)}{\pi(\hat{p}^1_{\Pi_n})} + \frac{\sum_{\nu^n \in \mathrm{C}} \pi(\nu^n)}{\pi(\hat{p}^1_{\Pi_n})})} \tag{6}$$

MaxProb/MaxEnt Thm implies that the first ratio in the denominator converges to $\mathrm{k} - 1$. By the Lemma, the second ratio in the denominator of (6) converges to zero as n goes to infinity. The second term in the nominator as well goes to zero as $n \to \infty$ (to see this, express the joint probability $\pi(X_1 = x_1, \ldots, X_t = x_t, \nu^n)$ as $\pi(X_1 = x_1, \ldots, X_t = x_t | \nu^n)\pi(\nu^n)$ and employ the Lemma).

Then, MaxProb/MaxEnt Thm implies, that for $n \to \infty$ the RHS of (6) becomes equal to $1/\mathrm{k} \sum_{j=1}^{\mathrm{k}} \pi(X_1 = x_1, \ldots, X_t = x_t | \hat{p}^j)$. Finally, invoke Csiszár's 'urn argument' (cf. [7]) to conclude that the asymptotic form of the RHS of (6) is $1/\mathrm{k} \sum_{j=1}^{\mathrm{k}} \prod_{l=1}^{t} \hat{p}^j_{X_l}$. □

11.4 Rational *I*-projections

Types can concentrate, in some sense, on rational I-projection $\hat{p} \in \mathrm{Q}^m$ even though the I-projection is isolated point of the set Π. The following Example illustrates the concentration.

Example 5. Consider $\Pi = \{p, \boldsymbol{p}\}$, where $p = [n_1/n_0, \ldots, n_m/n_0]$, $\boldsymbol{p} = [\boldsymbol{n}_1/n_0, \ldots, \boldsymbol{n}_m/n_0]$, $n_0 \in \mathrm{N}$. For $n \neq kn_0$, $k \in \mathrm{N}$ the set Π_n is empty; otherwise it contains p and $\boldsymbol{p}$. In this case, concentration of types on μ-projection is a direct consequence of the next two Lemmas. The I-variant of the concentration then arises from MaxProb/MaxEnt Thm.

Lemma 1. *Let* ν^n, $\boldsymbol{\nu}^n$ *be two n-types. Let* $\delta = \nu^n - \boldsymbol{\nu}^n$. *Let K denote the non-negative elements of* $n\delta$, *L the absolute value of negative elements of* $n\delta$. *Let* $c = \prod_- \boldsymbol{n}_i^{L_i} / \prod_+ \boldsymbol{n}_i^{K_i}$, *where the subscript* $-$, $+$ *indicates that the index i goes through the elements of K, L, respectively. Then* $\frac{\Gamma(k\nu^{kn})}{\Gamma(k\boldsymbol{\nu}^{kn})} < c^k$, *for any* $k \in \mathrm{N}$.

Proof. $\frac{\Gamma(k\nu^{kn})}{\Gamma(k\boldsymbol{\nu}^{kn})} = \frac{\prod_-(k(\boldsymbol{n}_i - L_i)+1)\cdots k\boldsymbol{n}_i}{\prod_+(k\boldsymbol{n}_i+1)\cdots k(\boldsymbol{n}_i+K_i)}$. So $\frac{\Gamma(k\nu^{kn})}{\Gamma(k\boldsymbol{\nu}^{kn})} \leq \frac{\prod_- \boldsymbol{n}_i^{kL_i}}{\prod_+ \boldsymbol{n}_i^{kK_i}}$, which is just c^k. □

Lemma 2. *Let types* ν^n, $\boldsymbol{\nu}^n$ *be such that* $\pi(\boldsymbol{\nu}^n; q) < \pi(\nu^n; q)$. *Then* $\frac{\pi(k\nu^{kn}; q)}{\pi(k\boldsymbol{\nu}^{kn}; q)} \to 0$ *as* $k \to \infty$.

Proof. By the assumption, $\prod q_i^{n_i - \boldsymbol{n}_i} < \frac{\Gamma(\tilde{\nu}^n)}{\Gamma(\nu^n)}$. The gamma-ratio is, by the Lemma 1, smaller or equal to c, as defined at the Lemma. Thus, $\prod q_i^{n_i - \boldsymbol{n}_i} = \gamma c$, where $\gamma \in [0, 1)$, $\gamma \in \mathrm{R}$. By Lemma 1, for any $k \in \mathrm{Z}$, $\frac{\pi(k\nu^{kn}; q)}{\pi(k\boldsymbol{\nu}^{kn}; q)} \leq (1/c)^k \prod q_i^{k(n_i - \tilde{n}_i)}$. The RHS of the inequality, γ^k, goes for $k \to \infty$ to zero, which completes the proof. □

In this case, if Π admits several rational I/μ-projections, then clearly, types equi-concentrate on them.

References

[1] Bártfai, P. "On a conditional limit theorem," *Progress in Statistics* **1**, 85-91, NH, Amsterdam, 1974.

[2] Chentsov, N. N., "Nonsymmetric distance measure of probability distributions, entropy and the Pythagoras Theorem," *Matem. Zametki*, **4**, No. 3, 323-332, 1968. (In Russian).

[3] Cohen, E. G. D., "Some recent advances in classical statistical mechanics," in LNP 597, P. Garbaczewski and R. Olkiewicz (eds), 7-33, Springer, Berlin, 2002.

[4] Cover, T. and Thomas, J., *Elements of Information Theory*, Wiley, NY, 1991.

[5] Csiszár, I., "Sanov property, generalized I-projection and a conditional limit theorem," *Ann. Probab.*, **12**, 768-793, 1984.

[6] Csiszár, I., "MaxEnt, mathematics and information theory," *Maximum Entropy and Bayesian Methods*, K. M. Hanson and R. N. Silver (eds.), pp. 35-50, Kluwer, 1996.

[7] Csiszár, I., "The method of types," *IEEE IT*, **44**, No. 6, 2505-2523, 1998.

[8] Dembo A. and Zeitouni O., *Large Deviations Techniques and Applications*, Springer, NY, 1998.

[9] Ellis, R. S., "The theory of large devitions: from Boltzmann's 1877 calcualtion to equilibrium macrostates in 2D turbulence," *Physica D*, 106-136, 1999.

[10] Golan, A., Judge, G. and Miller, D., *Maximum Entropy Econometrics. Robust Estimation with Limitted Data*, Wiley, NY, 1996.

[11] Grendár, M., Jr. and Grendár, M., "What is the question that MaxEnt answers? A probabilistic interpretation", in *Bayesian inference and Maximum Entropy methods in sicence and engineering*, A. Mohammad-Djafari (ed.), pp. 83-94, AIP, Melville, 2001. Also at arXiv:math-ph/0009020.

[12] Grendár, M., Jr. and Grendár, M., "Frequency moments inverse problem and Maximum (Shannon-Kullback vs. Renyi-Tsallis) Entropy," Preprint of IM PAN, Aug. 2002. Precursor of [15].

[13] Grendár, M., Jr. and Grendár, M., "Asymptotic equiprobability of I-projections," *Acta U. Belii Ser. Math.*, **10**, 3-8, 2003.

[14] Grendár, M., Jr. and Grendár, M., "Maximum Probability and Maximum Entropy methods: Bayesian interpretation," in *Bayesian inference and Maximum Entropy methods in sicence and engineering*, G. Erickson and Y. Zhai (eds.), pp. 490-494, AIP, Melville, 2004. Also at arXiv:physics/0308005.

[15] Grendár, M., Jr. and Grendár, M., "Maximum Entropy method with non-linear moment constraints: challenges," in *Bayesian inference and Maximum Entropy methods in sicence and engineering*, G. Erickson and Y. Zhai (eds.), pp. 97-109, AIP, Melville, 2004. Also at arXiv:physics/0308006.

[16] Grendár, M., "Asymptotic identity of μ-projections and I-projections," Acta U. Belii Ser. Math., **11**, pp. 3-7, 2004.

[17] Grendár, M., "Determination of constrained modes of a multinomial distribution," in *CompStat 2004*, J. Antoch (ed.), pp. 1109–1114, Springer, 2004.

[18] Grendár, M., "Conditioning by rare sources," to appear at Acta U. Belii Ser. Math., **12**, 2005.

[19] Grendár, M. and Judge, G., "Large deviations theory and empirical estimator choice," preprint ARE Berkeley, Jan 2006.

[20] Grendár, M., "L-divergence consistency for a discrete prior," working paper, 2006.

[21] Groeneboom, P., Oosterhoff, J. and Ruymgaart, F. H., "Large deviation theorems for empirical probability measures," *Ann. Probab.*, **7**, 553-586, 1979.

[22] Gzyl, H., "Ill-posed linear inverse problems and Maximum Entropy in the Mean," *Acta Cient. Venezolana*, **53**, 74-93, 2002.

[23] Harremoës, P., "Information topologies with applications," to appear at *Bolyai Studies*.

[24] Jaynes, E. T., *Papers on Probability, Statistics and Statistical Physics*, R. D. Rosenkratz (ed.), D. Reidel, Dordrecht, 1979.

[25] Jaynes, E. T., *Probability Theory: The Logic of Science,* G. L. Bretthorst (ed.), CUP, NY, 2003.

[26] Jizba, P. and Arimitsu, T., "The world according to Renyi: thermodynamics of fractal systems," arXiv:cond-mat/0108184, Aug 2001.

[27] La Cour, B. R. and Schieve, W. C., "Tsallis maximum entropy principle and the law of large numbers," *Phys. Rev. E*, **62**, 7494-7496, 2002. Also at arXiv:cond-mat/0009216.

[28] Lanford, O. E., "Entropy and equilibrium states in classical statistical mechanics," Statistical Mechanics and Mathematical Problems, *Lecture Notes in Physics* **20**, pp. 1-113, Springer, Berlin, 1973.

[29] Leonard, Ch. and Najim, J., "An extension of Sanov's Theorem. Application to the Gibbs Conditioning Principle," *Bernoulli*, **8**, 6, 721-743, 2002.

[30] Lewis, J. T., Pfister, C.-E. and Sullivan, W. G., "Entropy, concentration of probability and conditional theorems," *Markov Proc. Rel. Field.,* **1**, pp. 319–386, 1995.

[31] Mittelhammer, R., Judge, G. and Miller, D., *Econometric Foundations*, CUP, Cambridge, 2000.

[32] Owen A. B., *Empirical Likelihood*, Chapman-Hall/CRC, 2001.

[33] Romera, E., Angulo, J. C. and Dehesa, J. S., "Reconstruction of a density from its entropic moments," *Bayesian Inference and Maximum Entropy Methods*, R. L. Fry (ed.), pp. 449-457, AIP, 2002.

[34] Sanov, I. N., "On the probability of large deviations of random variables," *Mat. Sbornik,* **42**, pp. 11–44, 1957. (In Russian).

[35] Solana-Ortega, A., personal communication, Aug 2003.

[36] Topsøe, F., Information theoretical optimization techniques, *Kybernetika*, **15**, pp. 8-27, 1979.

[37] Tsallis, C. and Brigatti, E., "Nonextensive statistical mechanics: a brief introduction," arXiv:condmat/0305606, May 2003.

[38] van Campenhout, J. M. and Cover, T. M., "Maximum entropy and conditional probability," *IEEE IT*, **27**, 483-489, 1981.

[39] Vasicek, O. A., "A conditional law of large numbers," *Ann. Probab.*, **8**, 142-147, 1980.

[40] Vincze, I., "On the maximum probability principle in statistical physics," *Coll. Math. Soc. J. Bolyai*, **9**, European Meeting of Statisticians, Budapest, 869-893, 1972.

[41] Zabel, S., "Rates of convergence for conditional expectations," *Ann. Probab*, **8**, 928-941, 1980.

In: Focus on Probability Theory
Editor: Louis R. Velle, pp. 91-108
ISBN 1-59454-474-3

Chapter 5

New Product Theorems in Markov Networks*

G. Sh. Tsitsiashvili † **and M. A.Osipova** ‡
Institute of Applied Mathematics
Far Eastern Branch of Russian Academy Sciences
690041, Vladivostok, Radio str. 7, IAM FEB RAS

Abstract

In this paper new product theorems of queuing theory have been proved. These theorems give accuracy analytical formulas to calculate stationary distributions in some queuing networks with complicated structures and protocols. Three directions of product theorems development have been considered in this paper. All these directions are closely connected with systems of mobile telephone connection [1].
First direction is product theorems in queuing networks with prohibitions. Such networks appear for example in queuing systems with retrial queues. A consideration of queuing network with prohibitions demands to decompose a system of the network motion equations in an accordance with a graph of permissible transitions.
Second direction is a consideration of queuing network behavior in a random environment and its adaptation to this environment. For this aim a technique of conditional product stationary distributions is developed. These directions of product theorems development are closely connected with attendant transportation problems. Special algorithms of a search of permissible solutions of these problems have been constructed.
Last direction is a consideration of queuing networks with switching centers in which a route of the customer doesn't change. Special properties of Poisson distribution allow to conserve product form of stationary distributions in queuing network with such centers.

1. Product Theorems in Queuing Networks with Prohibitions

In classical product theorems [3], [6] it is possible to recognize an analogy between opened and closed queuing networks. A closed queuing network may be represented as an opened

*The paper is supported by RFBR, project 03-01-00512.
†E-mail address: guram@iam.dvo.ru
‡E-mail address: mao1975@list.ru

network in which both arrivals to the network and departures from the network are prohibited. These prohibitions do not prevent to represent stationary distributions of customer numbers at different nodes in the product form. Is it possible to widen a set of such prohibitions with the product form of stationary distributions? This problem occurs from modern applications for example retrial queues [9], [10], [11].

In this section some new product theorems for opened and closed queuing networks with finite number of states have been proved. An each network is characterized by a graph with states in its nodes and positive transition intensities in its edges. For different graphs and systems of motion equations stationary product distributions of Markov processes, which describe queuing networks with different prohibited transitions, have been obtained. Calculation algorithm of route matrices, corresponding to modified systems of motion equations, has been constructed.

1.1. Formulation of Main Results

Consider an opened queuing network G with m nodes (one-server queuing systems), input intensity $\lambda > 0$, serving intensities $\mu_1 > 0, \ldots, \mu_m > 0$, and route matrix $\Theta = (\theta_{ij})_{k,i=0}^{m}$:

$$\theta_{00} = 0,\ \ 0 \leq \theta_{ki} \leq 1,\ \ (k,i) \neq (0,0),\ \ \sum_{i=0}^{m} \theta_{ki} = 1,\ \ 0 \leq k,i \leq m.$$

Here θ_{ki} is a probability of customer transition from i-th node to j-th node after its serving in i-th node (where 0-th node is an external source or external space).

Denote $\mathbf{e}_1 = (1,0,\ldots,0)$, $\mathbf{e}_2 = (0,1,0,\ldots,0), \ldots, \mathbf{e}_m = (0,\ldots,0,1)$ basic vectors in the linear space E^m and let $Z^m = \{\mathbf{N} = (n_1,\ldots,n_m),\ n_1 \geq 0,\ldots,n_m \geq 0\}$. Suppose that $L(\mathbf{N},\mathbf{J})$, $\mathbf{J} = (j_1,\ldots,j_m)$, $j_1, j_2, \ldots, j_m \geq 0$, are transition intensities of discrete Markov process $\mathbf{N}(t)$ with states in Z^m, which describe numbers of customers in nodes of opened Jackson network with route matrix Θ and input intensity λ. Positive transition intensities of Markov process $\mathbf{N}(t)$ are the following

$$L(\mathbf{N},\mathbf{N}+\mathbf{e}_k) = \lambda\theta_{0k},\ \ \mathbf{N} \in Z^m,\ \ L(\mathbf{N},\mathbf{N}-\mathbf{e}_k) = \mu_k\theta_{k0},\ \ \mathbf{N} \in Z^m,\ \ n_k > 0, \tag{1.1}$$

$$L(\mathbf{N},\mathbf{N}-\mathbf{e}_k+\mathbf{e}_i) = \mu_k\theta_{ki},\ \ \mathbf{N} \in Z^m,\ \ n_k > 0,\ \ 1 \leq k \neq i \leq m. \tag{1.2}$$

To describe the network G with prohibitions consider non-oriented graph Γ with nodes from states set $L \subseteq Z^m$ and with edges represented in the form

$$[\mathbf{N},\mathbf{N}+\mathbf{e}_k],\ \ \mathbf{N} \in Z^m,\ \ 1 \leq k \leq m, \tag{1.3}$$

$$[\mathbf{N},\mathbf{N}-\mathbf{e}_k+\mathbf{e}_i],\ \ \mathbf{N} \in Z^m,\ \ n_k > 0;\ \ 1 \leq k \neq i \leq m. \tag{1.4}$$

If (1.3) edge belongs to the graph Γ then there is an allowance for a customer transition from k-th node to 0-th node and from 0-th node to k-th node, $1 \leq k \leq m$. If (1.4) edge belongs to the graph Γ then there is an allowance for customer served in i-th node to transit to k-th node and vica versa. If some of (1.3), (1.4) edges does not belong to the graph Γ then appropriate transitions are prohibited in both sides.

If (1.3) (or (1.4)) edge is prohibited then a customer arrived in k-th node from i-th node returns to i-th node and vica versa, $0 \leq k \neq i \leq m$, and so transition intensity (1.1) (or transition intensity (1.2)) becomes equal to 0. Such allowances and prohibitions define a protocol in the network G for example retrial queues protocol.

The network G with states set $L = Z^m$ and all possible (1.3), (1.4) edges (infinite queues in network nodes) is opened Jackson network. If the route matrix is indivisible:

$$\forall i, j \in \{0, 1, ..., m\} \; \exists \; i_1, i_2, \ldots i_r \in \{1, ..., m\} \; : \theta_{ii_1} > 0, \; \theta_{i_1 i_2} > 0, \ldots \theta_{i_r j} > 0,$$

then the system of motion equations

$$(\lambda_1, \ldots, \lambda_m) = (\lambda, \lambda_1, \ldots, \lambda_m)\Theta_0 \tag{1.5}$$

has a unique solution and $\lambda_1 \geq 0, \ldots, \lambda_m \geq 0$. Here the matrix Θ_0 is obtained from the matrix Θ by a removal of 0-th column. If $\lambda_1 < \mu_1, \ldots, \lambda_m < \mu_m$ then the discrete Markov process $\mathbf{N}(t)$, $t \geq 0$, is ergodic [7] and its limit distribution [3] is calculated by

$$P(\mathbf{N}) = C\Phi(\mathbf{N}), \; \Phi(\mathbf{N}) = \prod_{i=1}^{m} \left(\frac{\lambda_i}{\mu_i}\right)^{n_i}, \; C = \left[\sum_{\mathbf{N} \in L} \Phi(\mathbf{N})\right]^{-1}, \; \mathbf{N} \in L. \tag{1.6}$$

Theorem 1. *Suppose that the route matrix Θ of opened network satisfies a condition*

$$0 < \theta_{0k}, \theta_{k0} < 1, \; 1 \leq k \leq m, \tag{1.7}$$

and a vector $(\lambda_1, \ldots, \lambda_m)$ is a solution of the system (1.5). Let L_0 is finite subset of Z^m so that the graph Γ consisting of edges $\{[\mathbf{N}, \mathbf{N} + \mathbf{e}_i], \; [\mathbf{N} + \mathbf{e}_i, \mathbf{N} + \mathbf{e}_k] : \; \mathbf{N} \in L_0, \; 1 \leq i \neq k \leq m\}$ is connected. Then the discrete Markov process $\mathbf{N}(t)$ is ergodic and its limit distribution $P(\mathbf{N})$, $\mathbf{N} \in L = L_0 \bigcup \{\mathbf{N} + \mathbf{e}_i : \mathbf{N} \in L_0, \; 1 \leq i \leq m\}$ is calculated by (1.6).

Following sets $\{\mathbf{N} \in Z^m : \; \sum_{k=1}^m n_k < M\}$, $\{\mathbf{N} \in Z^m : \; M' < \sum_{k=1}^m n_k < M\}$, $\{\mathbf{N} \in Z^m : M'_k \leq n_k < M_k, \; 1 \leq k \leq m\}$, $0 \leq M', M, M'_1, M_1, \ldots, M'_m, M_m < \infty$ satisfy conditions of this theorem.

Theorem 2. *Suppose that the conditions (1.7) are true and L is finite subset of Z^m and Γ is connected graph with nodes which create a set L and with (1.3), (1.4) edges. If elements of the matrix Θ and numbers $\lambda_1 > 0, \ldots, \lambda_m > 0$, satisfy equalities*

$$\lambda_k \theta_{k0} = \lambda \theta_{0k}, \; 1 \leq k \leq m, \tag{1.8}$$

$$\lambda_i \theta_{ik} = \lambda_k \theta_{ki}, \; 1 \leq k \neq i \leq m, \tag{1.9}$$

then the discrete Markov process $\mathbf{N}(t)$ is ergodic and its limit distribution $P(\mathbf{N})$, $\mathbf{N} \in L$ is calculated by (1.6).

Consider now closed Jackson network G' with states set $L = \{\mathbf{N} \in Z^m, \sum_{i=1}^m n_i = M\}$ and all (1.4) edges. Here M is the total number of customers in the set G'. The set G' is described by discrete Markov process $\mathbf{N}(t)$, $t \geq 0$, with states set L and transition intensities (1.2). If route matrix $\Theta' = \left(\theta'_{ki}\right)_{k,i=1}^m$ of the set G' satisfies the condition

$$0 < \theta'_{ki} < 1, \; 1 \leq k, i \leq m \tag{1.10}$$

then appropriate system of motion equations [6]

$$(\lambda_1,\dots,\lambda_m)=(\lambda_1,\dots,\lambda_m)\Theta' \tag{1.11}$$

has infinite number of solutions $\lambda_1,\dots,\lambda_m$ with nonnegative components. For any natural M and for any solution of the system (1.11) with nonnegative components the formula (1.6) is true.

Theorem 3. *Suppose that route matrix Θ' of closed network satisfies the condition (1.10) and a vector $(\lambda_1,\dots,\lambda_m)$ is a solution of the system (1.11) with nonnegative components. Suppose that $\mathcal{L}_0$ is finite subset of Z^m so that the graph Γ consisting of edges $\{[\mathbf{N}+\mathbf{e}_i,\mathbf{N}+\mathbf{e}_k]:\ \mathbf{N}\in\mathcal{L}_0,\ 1\le i\ne k\le m\}$ is connected. Then discrete Markov process $\mathbf{N}(t)$ is ergodic and its limit distribution $P(\mathbf{N}),\mathbf{N}\in\mathcal{L}=\{\mathbf{N}+\mathbf{e}_i:\mathbf{N}\in\mathcal{L}_0,\ 1\le i\le m\}$ is calculated by (1.6).*

Theorem 4. *Suppose that $\mathcal{L}$ is finite subset of Z^m and Γ is connected graph with nodes, which create a set $\mathcal{L}$, and with (1.4) edges. If elements of the matrix Θ' and numbers $\lambda_1>0,\dots,\lambda_m>0$ satisfy equalities*

$$\lambda_i\theta'_{ik}=\lambda_k\theta'_{ki},\ \ 1\le k,i\le m, \tag{1.12}$$

then discrete Markov process $\mathbf{N}(t)$ is ergodic and its limit distribution $P(\mathbf{N}),\ \mathbf{N}\in\mathcal{L}$, is calculated by (1.6).

1.2. Proof of Main Results

Proof of the theorem 1. Denote

$$F_0(\mathbf{N})=\{\mathbf{N}+\mathbf{e}_i,\ 1\le i\le m\},$$

$$F_i(\mathbf{N})=\{\mathbf{N}-\mathbf{e}_i,\ \mathbf{N}-\mathbf{e}_i+\mathbf{e}_j,\ 1\le j\le m,\ j\ne i\},\ 1\le i\le m,\ \mathbf{N}\in Z^m.$$

It is clear that

$$F_i(\mathbf{N})\bigcap F_j(\mathbf{N})=\oslash. \tag{1.13}$$

As conditions of the theorem 1 are true so in the graph Γ for any $\mathbf{N}\in\mathcal{L}$ there is a set of indexes $I(\mathbf{N})\subseteq\{0,\dots,m\}$ so that a set of nodes $S(\mathbf{N})$ connected with the node $\mathbf{N}$ by single edges is represented as follows

$$S(\mathbf{N})=\bigcup_{i\in I(\mathbf{N})}F_i(\mathbf{N}). \tag{1.14}$$

From (1.1), (1.2), (1.5) obtain

$$\sum_{\mathbf{J}\in F_i(\mathbf{N})}[\Phi(\mathbf{N})L(\mathbf{N},\mathbf{J})-L(\mathbf{J},\mathbf{N})\Phi(\mathbf{J})]=0,\ 0\le i\le m,\ \mathbf{N}\in\mathcal{L}. \tag{1.15}$$

From (1.15) obtain for $\mathbf{N}\in\mathcal{L}$

$$\sum_{\mathbf{J}\in\mathcal{L}}[\Phi(\mathbf{N})L(\mathbf{N},\mathbf{J})-L(\mathbf{J},\mathbf{N})\Phi(\mathbf{J})]=\sum_{\mathbf{J}\in S(\mathbf{N})}[\Phi(\mathbf{N})L(\mathbf{N},\mathbf{J})-L(\mathbf{J},\mathbf{N})\Phi(\mathbf{J})]=0. \tag{1.16}$$

The graph Γ is connected and the conditions (1.7) and the equalities (1.16) are true and transition intensities $L(\mathbf{N},\mathbf{J})$ are bounded by $\lambda+\mu_1+\ldots+\mu_m$. So the process $\mathbf{N}(t)$, describing an opened queuing network with states set $\mathcal{L}$ and with the graph Γ, satisfies well known ergodicity theorem for Markov processes [7]. So it is ergodic and its limit distribution $P(\mathbf{N})$ satisfies the formula (1.6). The theorem 1 is proved.
Proof of the theorem 3. It repeats the proof of the theorem 1 practically word by word. Small changes touch only the inclusion $I(\mathbf{N}) \subseteq \{0,\ldots,m\}$, which is replaced by the inclusion $I(\mathbf{N}) \subseteq \{1,\ldots,m\}$. And in (1.15) the inequality $0 \leq i \leq m$ is replaced by the inequality $1 \leq i \leq m$.
Proof of the theorem 2. The equalities (1.6), (1.8), (1.9) lead to

$$\Phi(\mathbf{N})\lambda\theta_{0k} - \Phi(\mathbf{N}+\mathbf{e_k})\mu_k\theta_{k0} = 0, \;\; \mathbf{N},\, \mathbf{N}+\mathbf{e_k} \in \mathcal{L}, \;\; 1 \leq k \leq m, \tag{1.17}$$

$$\Phi(\mathbf{N})\mu_i\theta_{ik} - \Phi(\mathbf{N}+\mathbf{e}_k-\mathbf{e}_i)\mu_k\theta_{ki} = 0, \;\; \mathbf{N},\mathbf{N}+\mathbf{e}_k-\mathbf{e}_i \in \mathcal{L}, \;\; 1 \leq k \neq i \leq m. \tag{1.18}$$

As the formulas (1.1), (1.2), (1.17), (1.18) are true so the function $\Phi(\mathbf{N})$ satisfies the equations (1.16). An end of the theorem 2 proof repeats the end of the theorem 1 proof. The theorem 2 is proved.
Proof of the theorem 4. The equalities (1.6), (1.12) lead to the formulas (1.18) so the function $\Phi(\mathbf{N})$ satisfies the equations (1.16). An end of the theorem 4 proof repeats the end of the theorem 1 proof. The theorem 4 is proved.

Remark 1. *If in the conditions of the theorems 1 ñ 4 $\lambda_1 < \mu_1, \ldots, \lambda_m < \mu_m$, then in these theorems, the set $\mathcal{L}$ may be infinite.*

Remark 2. *In the theorem 2 (the theorem 4) new results are obtained not by special choice of the graph Γ (it is arbitrary here) but by a replacement of Jackson motion equations (1.5) (Gordon-Newell motion equations(1.11)) by modified motion equations (1.8), (1.9) (modified motion equations (1.12)).*

Remark 3. *It is clear from the proof that the theorem 1 is true for any connected graph Γ satisfying the following condition. For any node $\mathbf{N}$ of the graph Γ there is a set of indexes $I(\mathbf{N}) \subseteq \{0,\ldots,m\}$ so that $S(\mathbf{N})$ may be represented in the form (1.14). It is easy to prove that the conditions of the theorem 1 describe all graphs Γ which satisfy this condition. Analogous statement is true for the theorem 3.*

1.3. Algorithms of Route Matrix Construction

Opened Network. Suppose that $\lambda > 0,\ \theta_{0k} > 0,\ 1 > \theta_{k0} > 0,\ 1 \leq k \leq m$, are fixed. Using the formulas (1.8) it is possible to define $\lambda_1,\ldots,\lambda_m$ and θ_{ki}, $1 \leq k,i \leq m$, which are solutions of the following problem.
Denote

$$\Lambda_k = \lambda_k(1-\theta_{k0}), \tag{1.19}$$

then from (1.8)

$$\Lambda_k = \frac{\lambda\theta_{0k}(1-\theta_{k0})}{\theta_{k0}}, \;\; 1 \leq k \leq m. \tag{1.20}$$

Let

$$\pi_{ki} = \theta_{ki}/(1-\theta_{k0}), \tag{1.21}$$

so

$$\sum_{i=1}^{m} \pi_{ki} = 1, \ \ 1 \le k \le m. \tag{1.22}$$

Denote

$$C_{ki} = \Lambda_k \pi_{ki}, \ \ 1 \le k, i \le m. \tag{1.23}$$

As the formulas (1.9), (1.22) are true so the matrix $C = (C_{ki})_{k,i=1}^{m}$ is a permissible solution of the transportation problem

$$\sum_{i=1}^{m} C_{ki} = \Lambda_k, \ \ C_{ki} = C_{ik} > 0, \ \ 1 \le k, i \le m. \tag{1.24}$$

So using fixed numbers $\lambda > 0$, $\theta_{0k} > 0$, $1 > \theta_{k0} > 0$, $1 \le k \le m$, find by the formulas (1.19) Λ_k, $1 \le k \le m$. Then using Λ_k, $1 \le k \le m$, define a permissible solution of the transportation problem (1.24) and from the formulas (1.21), (1.23) find π_{ki}, θ_{ki}, $1 \le k, i \le m$.

Consider now a search of permissible solutions of the transportation problem (1.24). Renumber nodes $1, \ldots, m$ so that $\Lambda_1 \le \Lambda_2, \ldots, \Lambda_1 \le \Lambda_m$. Choose $C_{11}, \ldots, C_{1m}$ from the conditions

$$\sum_{k=1}^{m} C_{1k} = \Lambda_1, \ \ C_{11} > 0, \ldots, C_{1m} > 0$$

and let

$$C_{k1} = C_{1k}, \ \ k = 1, \ldots, m.$$

Redefine now $\Lambda_2, \ldots, \Lambda_m$ by the following formulas:

$$\Lambda_2 := \Lambda_2 - C_{21}, \ldots, \Lambda_m := \Lambda_m - C_{m1}.$$

Now it is necessary to define C_{ki}, $2 \le k, i \le m$, from

$$\sum_{i=2}^{m} C_{ki} = \Lambda_k, \ \ C_{ki} = C_{ik} > 0, \ \ 2 \le k, i \le m. \tag{1.25}$$

Apply the same procedure but to the problem (1.25) and so on. Resulting algorithm is a modification of well known North-West Angle algorithm [4] of transportation problem (1.24) solution. The only difference is in additional condition that the matrix C is symmetric.

Closed Network. Suppose that intensities $\lambda_1 > 0, \ldots, \lambda_m > 0$ are fixed. Denote

$$C'_{ki} = \lambda_k \theta'_{ki}, \ \ 1 \le k, i \le m. \tag{1.26}$$

As the formula (1.12) and equalities

$$\sum_{i=1}^{m} \theta'_{ki} = 1, \ \ k = 1, \ldots, m,$$

are true so the matrix $C' = (C'_{ki})_{k,i=1}^{m}$ is permissible solution of the transportation problem

$$\sum_{i=1}^{m} C'_{ki} = \lambda_k, \; C'_{ki} = C'_{ik} > 0, \; 1 \leq k, i \leq m. \tag{1.27}$$

So to define the matrix C' it is possible to use mentioned algorithm. Then using known matrix C' and fixed intensities $\lambda_1, \ldots, \lambda_m$ define by (1.26) the route matrix Θ'.

2. Random Control of Markov Process Parameter

In this section we consider Markov processes with randomly varying parameter (or in random environment). Suppose that finite horizon or stationary process distributions for different fixed parameters do not coincide. The problem is to choose rules of random parameter control so that resulted process distributions (finite horizon or stationary) equal to some probability mixtures of the process distributions with fixed parameters. This concept of the random parameter control has an analog with the concept of pure and mixed states in quantum theory or with the concept of pure and mixed strategies in the game theory.

If the processes distributions for different fixed parameters coincide with each other, similar problems have been considered for stationary distributions of queuing systems [8], [12] - [14]. Here we developed some ideas of the paper [1]. Main result of this paper is an existence of one-to-one transformation between a set of solutions of random control parameter problem and a set of solutions of some dual transportation problem.

2.1. Control of Finite Horizon Distributions in Markov Chains

Suppose that there are finite sets of states I, J and Markov chains $x^{(i)}(n)$, $n \geq 0$, $i \in I$, (Markov processes with discrete time) with transition probabilities matrices $||Q^{(i)}_{t,j}(n)||_{t,j \in J}$, which consist of positive elements, and distributions $B^{(i)}_j(n) = P(x^{(i)}(n) = j)$ defined by recurrent formulas

$$B^{(i)}_j(n+1) = \sum_{t \in J} B^{(i)}_t(n) Q^{(i)}_{t,j}(n), \; j \in J; \; i \in I; \; n \geq 0. \tag{2.28}$$

Fix arbitrary positive numbers A_i, $i \in I$, $\sum_{i \in I} A_i = 1$, and consider Markov chain z_n, $n \geq 0$, with distributions $\Psi_{i,j}(n) = P(z_n = (i,j))$, defined by initial conditions $\Psi_{i,j}(0) = A_i B^{(i)}_j(0) > 0$ and recurrent formulas

$$\Psi_{i,j}(n+1) = \sum_{k \in I, t \in J} \Psi_{k,t}(n) Q^{(k)}_{t,j}(n) P^{(j)}_{k,i}(n), \; i \in I, \; j \in J, \; n \geq 0. \tag{2.29}$$

Our purpose is to find the control $U = (P^{(j)}(n) = ||P^{(j)}_{k,i}(n)||_{k,i \in I}, \; j \in J, \; n \geq 0)$, which consists of transition probabilities matrices satisfying the equalities

$$\Psi_{i,j}(n) = A_i B^{(i)}_j(n), \; i \in I, \; j \in J, \; n \geq 0. \tag{2.30}$$

Denote $F^{(j)}_k(n) = A_k B^{(k)}_j(n+1) > 0$ and take

$$G^{(j)}_{k,i}(n) = F^{(j)}_k(n) P^{(j)}_{k,i}(n), \; i \in I, \; k \in I, \; j \in J, \; n \geq 0. \tag{2.31}$$

Theorem 5. *The formula (2.31) determines one-to-one transformation between the set of the controls U, which satisfies the condition (2.2.), and the set of the collections $V = (||G_{k,i}^{(j)}(n)||_{k,i\in I},\ j \in J,\ n \geq 0)$ of the matrices, which are solutions of the following transportation problems for $j \in J,\ n \geq 0$:*

$$\sum_{k\in I} G_{i,k}^{(j)}(n) = F_i^{(j)}(n),\ i \in I, \tag{2.32}$$

$$\sum_{k\in I} G_{k,i}^{(j)}(n) = F_i^{(j)}(n),\ i \in I. \tag{2.33}$$

Proof. Suppose that the control U satisfies the condition (2.2.). Then it is easy to prove that the formula (2.31) and the complete probability formula give the formula (2.32). Insert the formula (2.2.) into the formula (2.29) and using the formula (2.28) obtain for $i \in I,\ j \in J,\ n \geq 0$:

$$A_i B_j^{(i)}(n+1) = \sum_{k\in I,\, t\in J} A_k B_t^{(k)}(n) Q_{t,j}^{(k)} P_{k,i}^{(j)}(n) = \sum_{k\in I} A_k P_{k,i}^{(j)}(n) \sum_{t\in J} B_t^{(k)}(n) Q_{t,j}^{(k)} =$$

$$= \sum_{k\in I} A_k P_{k,i}^{(j)}(n) B_j^{(k)}(n+1) = \sum_{k\in I} A_k B_j^{(k)}(n+1) P_{k,i}^{(j)}(n). \tag{2.34}$$

The formulas (2.31), (2.34) lead to the formula (2.33). So the collection V for each $j \in J,\ n \geq 0$, is a solution of the transportation problem (2.32), (2.33).

Suppose now that the collection $V = (||G_{k,i}^{(j)}(n)||_{k,i\in I},\ j \in J,\ n \geq 0)$ consists of the matrices, which are solutions of the transportation problems (2.32), (2.33) for $j \in J,\ n \geq 0$. Define the matrices of the control $U = (P^{(j)}(n) = ||P_{k,i}^{(j)}(n)||_{k,i\in I},\ j \in J,\ n \geq 0)$ by the formula (2.31) and prove that these matrices are transition probability matrices (which consist of nonnegative elements whose sums along each line equal to 1) and satisfy the equalities (2.2.). It is obvious that the formulas (2.31), (2.32) guarantee that all matrices in the control U are transition probability matrices.

For $n = 0$ the equalities (2.2.) are true by a definition. Suppose by an induction that these equalities are true for some $n = m$ and prove them for $n = m + 1$. Using the formulas (2.28), (2.29) and the suggestion of the induction obtain:

$$\Psi_{i,j}(m+1) = \sum_{k\in I,\, t\in J} A_k B_t^{(k)}(m) Q_{t,j}^{(k)} P_{k,i}^{(j)}(m) = \sum_{k\in I} A_k P_{k,i}^{(j)}(m) \sum_{t\in J} B_t^{(k)}(m) Q_{t,j}^{(k)} =$$

$$= \sum_{k\in I} A_k P_{k,i}^{(j)}(m) B_j^{(k)}(m+1) = \sum_{k\in I} A_k B_j^{(k)}(m+1) P_{k,i}^{(j)}(m). \tag{2.35}$$

Insert the formulas (2.31), (2.33) into the formula (2.35) and obtain the equality

$$\Psi_{i,j}(m+1) = A_i B_j^{(i)}(m+1).$$

So the formula (2.2.) is true for $n = m + 1$. The theorem 5 is proved.

Remark 4. *If for some $k \in I,\ j \in J,\ n \geq 0$, the equality $F_k^{(j)}(n) = A_k B_j^{(k)}(n+1) = 0$ takes place then the solution of the transportation problem (2.32) (2.33) satisfies the conditions $G_{k,i}^{(j)}(n) = 0,\ i \in I$, and so elements $P_{k,i}^{(j)}(n) \geq 0,\ i \in I,\ \sum_{i\in I} P_{k,i}^{(j)}(n) = 1$, of the matrix $P^{(j)}(n)$ may be arbitrary.*

2.2. Control of Stationary Distributions in Markov Chains

Suppose that there are finite sets of states I, J and Markov chains $x^{(i)}(n)$, $n \geq 0$, $i \in I$, with transition probabilities matrices $||Q^{(i)}_{t,j}||_{t,j\in J}$, which satisfy the communicability condition (**B**), and distributions $B^{(i)}_j(n) = P(x^{(i)}(n) = j)$ defined by recurrent formulas

$$B^{(i)}_j(n+1) = \sum_{t\in J} B^{(i)}_t(n) Q^{(i)}_{t,j}, \quad j \in J, \quad i \in I, \quad n \geq 0. \tag{2.36}$$

(**B**). All states of the chains $x^{(i)}(n)$, $i \in I$, are communicable:

$$\forall j,\ t \in J\ \exists j_1, \ldots j_r \in J: \ Q^{(i)}_{j,j_1} > 0,\ Q^{(i)}_{j_1,j_2} > 0, \ldots, Q^{(i)}_{j_r,t} > 0.$$

Using well known ergodicity theorems is easy to show that the chains $x^{(i)}(n)$, $n \geq 0$, $i \in I$, are ergodic with the stationary distributions $B^{(i)}_j > 0$, $j \in J$, $i \in I$:

$$B^{(i)}_j = \sum_{t\in J} B^{(i)}_t Q^{(i)}_{t,j}, \quad j \in J, \quad i \in I. \tag{2.37}$$

Fix arbitrary positive numbers A_i, $i \in I$, $\sum_{i\in I} A_i = 1$, and consider Markov chain z_n, $n \geq 0$, with distributions $\Psi_{i,j}(n) = P(z_n = (i,j))$ defined by initial conditions $\Psi_{i,j}(0) = A_i B^{(i)}_j(0)$ and recurrent formulas

$$\Psi_{i,j}(n+1) = \sum_{k\in I,\, t\in J} \Psi_{k,t} Q^{(k)}_{t,j} P^{(j)}_{k,i}, \quad i \in I, \quad j \in J, \quad n \geq 0. \tag{2.38}$$

Here the matrices $P^{(j)} = ||P^{(j)}_{k,i}||_{k,i\in I}$, $j \in J$, consist of positive elements and so the Markov chain z_n, $n \geq 0$, is ergodic with the stationary distribution

$$\Psi_{i,j} = \sum_{k\in I,\, t\in J} \Psi_{k,t}(n) Q^{(k)}_{t,j} P^{(j)}_{k,i}, \quad i \in I, \quad j \in J. \tag{2.39}$$

Our purpose is to find the control $U_* = (P^{(j)} = ||P^{(j)}_{k,i}||_{k,i\in I}$, $j \in J)$ which consists of transition probabilities matrices with positive elements satisfying the equalities

$$\Psi_{i,j} = A_i B^{(i)}_j, \quad i \in I, \quad j \in J. \tag{2.40}$$

Denote $F^{(j)}_k = A_k B^{(k)}_j > 0$ and take

$$G^{(j)}_{k,i} = F^{(j)}_k P^{(j)}_{k,i}, \quad i \in I, \quad k \in I, \quad j \in J. \tag{2.41}$$

Theorem 6. *The formula (2.41) determines one-to-one transformation between the set of the controls U_*, which satisfies the condition (2.3.), and the set of the collections $V_* = (||G^{(j)}_{k,i}||_{k,i\in I},\ j \in J)$ of the matrices with positive elements, which are solutions of the following transportation problems for $j \in J$:*

$$\sum_{k\in I} G^{(j)}_{i,k} = F^{(j)}_i, \quad i \in I, \tag{2.42}$$

$$\sum_{k\in I} G^{(j)}_{k,i} = F^{(j)}_i, \quad i \in I. \tag{2.43}$$

Proof. Suppose that the control U_* satisfies the condition (2.3.). Then it is easy to prove that the formula (2.41) and the complete probability formula give the formula (2.42). Insert the formula (2.3.) into the formula (2.38) and using the formula (2.36) obtain for $i \in I, j \in J$:

$$A_i B_j^{(i)} = \sum_{k \in I, t \in J} A_k B_t^{(k)} Q_{t,j}^{(k)} P_{k,i}^{(j)} = \sum_{k \in I} A_k P_{k,i}^{(j)} \sum_{t \in J} B_t^{(k)} Q_{t,j}^{(k)} =$$

$$= \sum_{k \in I} A_k P_{k,i}^{(j)} B_j^{(k)} = \sum_{k \in I} A_k B_j^{(k)} P_{k,i}^{(j)}. \tag{2.44}$$

The formulas (2.41), (2.44) lead to the formula (2.43). So the collection V_* for each $j \in J$ is a solution of the transportation problem (2.42), (2.43).

Suppose now that the collection $V_* = (||G_{k,i}^{(j)}||_{k,\, i \in I},\ j \in J)$ consists of the matrices, which are solutions of the transportation problems (2.42), (2.43) for $j \in J$. Define the matrices of the control $U_* = (P^{(j)} = ||P_{k,i}^{(j)}||_{k,\, i \in I},\ j \in J)$ by the formula (2.41) and prove that these matrices are transition probability matrices and satisfy the equalities (2.3.). It is obvious that the formulas (2.41), (2.42) guarantee that all matrices in the control U_* are transition probability matrices.

Using the formulas (2.36), (2.38) obtain:

$$\Psi_{i,j} = \sum_{k \in I, t \in J} A_k B_t^{(k)} Q_{t,j}^{(k)} P_{k,i}^{(j)} = \sum_{k \in I} A_k P_{k,i}^{(j)} \sum_{t \in J} B_t^{(k)} Q_{t,j}^{(k)} =$$

$$= \sum_{k \in I} A_k P_{k,i}^{(j)} B_j^{(k)} = \sum_{k \in I} A_k B_j^{(k)} P_{k,i}^{(j)}. \tag{2.45}$$

Insert the formulas (2.41), (2.43) into the formula (2.45) and obtain the equality (2.3.). The theorem 6 is proved.

2.3. Control of Stationary Distributions in Markov Processes

Suppose that there are finite sets of states I, J and Markov processes $x^{(i)}(\tau)$, $\tau \geq 0$, $i \in I$, with transition intensities matrices $||\mu_{t,j}^{(i)}||_{t,\, j \in J}$ (which consist of nonnegative elements outside of diagonal and zero diagonal elements) satisfy the following conditions:
(A). There exist solutions $\overline{B}_j^{(i)} > 0$, $j \in J$; $i \in I$, of the systems

$$\overline{B}_j^{(i)} \sum_{t \in J} \mu_{j,t} = \sum_{t \in J} \overline{B}_t^{(i)} \mu_{t,j}, \quad \sum_{j \in J} \overline{B}_j^{(i)} = 1. \tag{2.46}$$

$(\mathbf{B}^*)$. All states of processes $x^{(i)}(\tau)$, $i \in I$, are communicable:

$$\forall j,\, t \in J\ \exists j_1, \ldots j_r \in J: \ \mu_{j,\, j_1}^{(i)} > 0,\ \mu_{j_1,\, j_2}^{(i)} > 0, \ldots, \mu_{j_r,\, t}^{(i)} > 0.$$

If the processes $x^{(i)}(\tau)$, $\tau \geq 0$, $i \in I$, satisfy the conditions **(A)**, $(\mathbf{B}^*)$ then they are ergodic [7] and have stationary distributions $\overline{B}_j^{(i)}$, $j \in J$, $i \in I$.

Consider Markov process $z(\tau),\ \tau \geq 0$, with the transition intensities matrix

$$||\gamma((i,j),(k,t))||,\ k \in I,\ j,t \in J,$$

defined by equalities

$$\gamma((i,j),(i,t)) = \mu^{(i)}_{j,t},\ t \in J,\ i \in I,$$

$$\gamma((i,j),(k,j)) = \lambda^{(j)}_{i,k},\ j \in J,\ i,k \in I, \tag{2.47}$$

$$\gamma((i,j),(k,t)) = 0, \qquad \text{else}.$$

Here the following condition is true.

(C). The matrices $||\lambda^{(j)}_{i,k}||_{i,\,k\in I}$, $j \in J$, consist of positive elements outside of their diagonals and zero elements on diagonals.

Then it is easy to prove that Markov process $z(\tau),\ \tau \geq 0$, satisfy the conditions **(A)**, $(\mathbf{B}^*)$ and so is ergodic. Denote by $\Psi_{i,\,j},\ i \in I,\ j \in J$, the stationary distribution of this process and fix arbitrary positive numbers $A_i,\ i \in I,\ \sum_{i\in I} A_i = 1$.

Our purpose is to find the control $\overline{U} = (\Lambda^{(j)} = ||\lambda^{(j)}_{k,\,i}||_{k,\,i\in I}\ ,\ \ j \in J)$, which consists of the matrices satisfying the condition **(C)** and the equalities

$$\Psi_{i,\,j} = A_i \overline{B}^{(i)}_j,\ i \in I,\ j \in J. \tag{2.48}$$

Denote $\overline{F}^{(j)}_k = A_k \overline{B}^{(k)}_j > 0$ and take

$$\overline{G}^{(j)}_{k,\,i} = \overline{F}^{(j)}_k \lambda^{(j)}_{k,\,i},\ i \in I,\ k \in I,\ j \in J. \tag{2.49}$$

Theorem 7. *The formula (2.49) determines one-to-one transformation between the set of the controls $\overline{U}$, which satisfies the condition (2.48), and the set of the collections $\overline{V} = (||\overline{G}^{(j)}_{k,\,i}||_{k,\,i\in I}$, $j \in J)$ of the matrices, which satisfy the condition* **(C)**, *and for some $D^{(j)}_i > 0,\ i \in I,\ j \in J$, are solutions of the following transportation problems:*

$$\sum_{k\in I} \overline{G}^{(j)}_{i,\,k} = D^{(j)}_i,\ i \in I, \tag{2.50}$$

$$\sum_{k\in I} \overline{G}^{(j)}_{k,\,i} = D^{(j)}_i,\ i \in I,\ j \in J. \tag{2.51}$$

Proof. Suppose that the control U_* satisfies the condition (2.48). Define $D^{(j)}_i > 0,\ i \in I,\ j \in J$, by the formulas (2.50) and prove that the formulas (2.51) are true. Using the formulas (2.48) obtain:

$$\sum_{k\in I,t\in J} A_i \overline{B}^{(i)}_j \gamma((i,j),(k,t)) - \sum_{k\in I,t\in J} A_k \overline{B}^{(i)}_t \gamma((k,t),(i,j)) = 0. \tag{2.52}$$

Insert the equalities (2.47) into last formula and obtain:

$$\sum_{k\in I} A_i \overline{B}^{(i)}_j \lambda^{(j)}_{i,k} + \sum_{t\in J} A_i \overline{B}^{(i)}_j \mu^{(i)}_{j,\,t} - \sum_{k\in I} A_k \overline{B}^{(k)}_j \lambda^{(j)}_{k,i} - \sum_{t\in J} A_i \overline{B}^{(i)}_t \mu^{(i)}_{t,\,j} = 0. \tag{2.53}$$

Using the formula (2.46) obtain from (2.53) the formula (2.51).

Suppose now that for some $D_i^{(j)} > 0$, $i \in I$, $j \in J$, the matrices $||\overline{G}_{k,i}^{(j)}||_{k,i\in I}$, $j \in J$ satisfy the condition **(C)** and the formulas (2.50), (2.51). Using the formulas (2.49) define the control $\overline{U} = (\Lambda^{(j)} = ||\lambda_{k,i}^{(j)}||_{k,i\in I}, \ j \in J)$. It is obvious that the matrices $\Lambda^{(j)}$, $j \in J$, satisfy the condition **(C)** and so the states of the Markov process $z(\tau)$, $\tau \geq 0$, are communicable from the viewpoint of the condition $(\mathbf{B}^*)$. The formulas (2.50), (2.51) lead to the equalities

$$\sum_{k\in I} A_i \overline{B}_j^{(i)} \lambda_{i,k}^{(j)} - \sum_{k\in I} A_k \overline{B}_j^{(k)} \lambda_{k,i}^{(j)} = 0. \qquad (2.54)$$

The formulas (2.46), (2.54) lead to the formula (2.53) and so to the formula (2.52). As a result obtain that the Markov process $z(\tau)$, $\tau \geq 0$, satisfies the type **(A)** condition and so is ergodic with the stationary distribution (2.48). The theorem 7 is proved.

Example: Consider the simplest queuing system $M|M|1|n$ with input intensity $a > 0$ and serving intensity $b > 0$, $\rho = a/b$. Then stationary distribution $\overline{B}_j$, $j = 0,\dots n$, of birth and death process, which describes a number of customers in this system, is the following:

$$\overline{B}_j = \frac{\rho^j(1-\rho)}{(1-\rho^{n+1})}, \quad 0 \leq j \leq n.$$

Suppose that there are two different vectors $(a^{(i)}, b^{(i)})$, $i \in I = (1,2)$, with positive components and define stationary distributions corresponding to these vectors:

$$\overline{B}_j^{(i)} = \frac{\rho_i^j(1-\rho_i)}{(1-\rho_i^{n+1})}, \quad 0 \leq j \leq n, \ \rho_i = \frac{a_i}{b_i}, \quad i = 1, 2.$$

Using the vectors $(a^{(i)}, b^{(i)})$, $i = 1,2$, it is simple to determine the transition intensities matrices $||\mu_{t,j}^{(i)}||_{t,j\in J}$, $i = 1,2$, $J = (0,\dots,n)$, of birth and death processes $x^{(i)}(\tau)$, $\tau \geq 0$, $i \in I$, corresponding these vectors.

Fix positive numbers D, A_1, A_2 so that $A_1 + A_2 = 1$ and take arbitrary positive numbers $\lambda_{1,2}^{(j)}$, $\lambda_{2,1}^{(j)}$ from the equalities

$$D = A_1 \overline{B}_j^{(1)} \lambda_{1,2}^{(j)} = A_2 \overline{B}_j^{(2)} \lambda_{2,1}^{(j)}, \quad j = 0,\dots,n.$$

So we obtain the matrices $||\lambda_{i,k}^{(j)}||_{i,k\in I}$, $j \in J$, which satisfy the conditions (2.50) (2.51) with $D_i^{(j)} = D$, $i \in I$, $j \in J$. Define the process $z(\tau)$, $\tau \geq 0$, which describes the queuing system $M|M|1|n$ with input and serving intensities, which change in accordance with the matrices $||\lambda_{i,k}^{(j)}||_{i,k\in I}$, $j \in J$. Then the stationary distribution $\overline{\Psi}_{i,j}$ of the process $z(\tau)$ satisfies the formula (2.48).

2.4. Algorithms of Matrices Generation

Markov processes with discrete time (chains). Consider now how to generate all matrices which are solutions of appropriate transportation problems. For the sake of simplicity consider an algorithm of a generation of the following transportation problem solution. Suppose that

$$a_1 > 0, \ldots, a_n > 0, \ N = (1, \ldots, n), \ M = (1, \ldots, n-1), \ n > 1, \ a = \sum_{r \in N} a_r,$$

and consider matrices $B = ||b_{p,q}||_{p,q \in N}$, which satisfy the conditions:

$$\sum_{r \in N} b_{p,r} = a_p, \ \sum_{r \in N} b_{r,q} = a_q, \ b_{p,q} > 0; \ p, q \in N. \tag{2.55}$$

Such transportation problems occurred in the subsections 2.1, 2.2 and always have solutions. Our algorithm of matrices B generation is based on the algorithm **(D)** of an auxiliary problem solution.

Suppose that

$$0 < d_1, \ldots, 0 < d_n, \ d = d_1 + \ldots + d_n, \ 0 < c < d$$

and seek for the vector $(c_1, \ldots, c_n)$ so that

$$\sum_{r \in N} c_r = c, \ 0 < c_r < d_r, \ r \in N. \tag{2.56}$$

(D). Suppose that $0 < w_1, \ldots, w_{n-1} < 1$, denote $C_0 = 0$, $C_r = c_1 + \ldots + c_r$, $D_0 = 0$, $D_r = d_1 + \ldots + d_r$, $1 \leq r \leq n$, and define components of the vector $(c_1, \ldots, c_n)$ using recurrent formulas:

$$c_k = w_k \max(0, c - d - C_{k-1} + D_{k-1} + d_k) + (1 - w_k) \min(d_k, c - C_{k-1}), \ 1 \leq k < n$$

$$c_n = c - C_{n-1}. \tag{2.57}$$

It is easy to prove that each vector $(c_1, \ldots, c_n)$, defined by the algorithm **(D)**, satisfies the conditions (2.56) and visa versa. Then each matrix B, satisfying the conditions (2.55), may be defined through the following algorithm.

(E). At first determine $d_1 = a_1, \ldots, d_n = a_n$. Then for $k = 1, \ldots, n-1$ take $c = a_k$ and determine the vector $(c_1, \ldots, c_n)$ using the algorithm **(D)** by k - th line $(w_{k,1}, \ldots, w_{k,n-1})$ of the matrix W. Put $b_{k,1} = c_1, \ldots, b_{k,n} = c_n$ and redetermine $d_1 := a_1 - c_1, \ldots, d_n := a_n - c_n$ and for $k < n-1$ begin $k+1$ - th step by the operation $k := k+1$. For $k = n-1$ this cycle is finished and the lowest line of the matrix B is determined by the equalities $b_{n,1} = d_1, \ldots, b_{n,n} = d_n$.

The algorithms **(D)**, **(E)** create some generalization of the well known North-West Angle algorithm, in which the formula (2.57) is to be replaced by the equalities $c_k = \min(d_k, c - C_{k-1})$, $1 \leq k < n$. Specifics of the algorithms **(D)**, **(E)** is that they allow to transform each matrix $W = ||w_{s,r}||_{s,r \in M}$, $0 < w_{s,r} < 1$, $s, r \in M$, into the matrix B, satisfying the conditions (2.55). This transformation is peace wise multi-linear with free member and is one-to-one from the set of all considered matrices W (from $(n-1)^2$ - fold direct product of the interval $(0,1)$) to the set of all matrices B, satisfying the condition (2.55). But if we need some concrete matrices it is possible to use for example the following simple formula:

$$b_{p,q} = \varepsilon a_p \delta_{p,q} + (1 - \varepsilon) \frac{a_p a_q}{a}, \ p, q \in N,$$

where $0 < \varepsilon \leq 1$, $\delta_{p,q}$ is Kroneker symbol.

Markov processes. The transportation problems from the subsection 2.3 demand another algorithms because their solutions are matrices with zero diagonal elements. Consider the set of all finite sets of matrices $\overline{V} = (||\overline{G}^{(j)}_{k,i}||_{k,i\in I},\ j \in J)$, which have zero diagonal elements and positive elements out of the diagonal and satisfy the conditions

$$\sum_{k\in I} \overline{G}^{(j)}_{i,k} = \sum_{k\in I} \overline{G}^{(j)}_{k,i},\ \ i \in I,\ \ j \in J. \tag{2.58}$$

Fix $j \in J$ and denote $||\overline{G}^{(j)}_{k,i}||_{k,i\in I} = ||g_{k,i}||_{k,i\in I} = \mathbf{g}$ and represent $\mathbf{g}$ in the following form: $\mathbf{g} = \mathbf{g^s} + \mathbf{g^a}$ where $\mathbf{g^s}$ is symmetric and $\mathbf{g^a}$ is antisymmetric components of the matrice $\mathbf{g}$. An algorythm of a building of arbitrary matrice $\mathbf{g}$, which satisfies the equalities (2.58), consists of the following steps. At first take arbitrary symmetric $\mathbf{g^s} = ||g^s_{i,k}||$ matrice with positive elements out of the diagonal and zero elements at the diagonal. Then choose arbitrary antisymmetric matrice $\mathbf{f^a} = ||f_{i,k}||$ so that

$$\sum_{k\in I} f_{i,k} = 0,\ \ f_{i,k} \neq 0,\ \ i \neq k. \tag{2.59}$$

To build the matrice $\mathbf{f^a}$ it is necessary to choose arbitrary positive numbers $f_{i,k}$, $i < k < m = |I|$ (here $|I|$ is the number of elements in finite set I) and calculate $f_{i,m}$ from (2.59). Define

$$F = \min_{i\neq k} \frac{g^s_{i,k}}{|f_{i,k}|}$$

and for arbitrary R, $|R| < F$, take

$$g^a_{i,k} = Rf_{i,k},\ \ g_{i,k} = g^s_{i,k} + g^a_{i,k}.$$

It is easy to prove that this algorithm gives general solution of the problem (2.58).

3. Product Theorems in Queuing Networks with Switching Centers

This section is devoted to a consideration of opened and closed networks with switching centers. The switching centers are queuing systems which have infinite number of identical servers with exponentially distributed serving times. If a customer moves from one serving center to another serving center it passes via some switching center. In this center a route of the customer doesn't change. Such networks origined in a model of mobile telephone system suggested by prof. D. Baum in [1]. The classical product theorems for BCMP queuing networks [5] can't be applied directly to calculate stationary distribution in the network with united switching centers. Nevertheless there is a possibility to make necessary calculations using classical product theorems and some properties of switching centers.

3.1. Opened Networks

Consider opened network J (partial case of BCMP queuing network). Suppose that it has Poisson input flow with the intensity λ and m multiserver centers and $i-$ th center has $r_i \leq \infty$ servers (each server has exponentially distributed serving times with intensity μ_i) and allows infinite queue, $i = 1, 2, ..., m$. Take external source like a center with the index 0. Determine a customer route in the network J by the route matrix $\Theta_J = (\theta_{ij})_{i,j=0}^{m}$, where θ_{ij} is a probability to transit from $i-$th serving center to $j-$th serving center, $\theta_{00} = 0$. Suppose that the route matrix is indivisible:

$$\forall i, j \in \{0, 1, ..., m\} \ \exists \ i_1, \ i_2, \ldots i_r \in \{1, ..., m\} \ : \ \theta_{ii_1} > 0, \ \theta_{i_1 i_2} > 0, \ldots \theta_{i_r j} > 0.$$

So there is a single solution $(\bar{\lambda}, \bar{\lambda}_1, \bar{\lambda}_2, \ldots \bar{\lambda}_m)$ of the system

$$\bar{\lambda} = \lambda, \ \ (\bar{\lambda}, \bar{\lambda}_1, \bar{\lambda}_2, ..., \bar{\lambda}_m) = (\bar{\lambda}, \bar{\lambda}_1, \bar{\lambda}_2, ..., \bar{\lambda}_m)\Theta_J.$$

Separate in the network J i-th serving center and consider it like isolated r_i-server queuing system with Poisson input flow, which has intensity $\bar{\lambda}_i$. Describe a number of customers in this isolated system by birth and death process $k_i(t)$. If

$$\rho_i = \frac{\bar{\lambda}_i}{r_i \mu_i} < 1, \tag{3.1}$$

then the process $k_i(t)$ is ergodic and has stationary distribution [7]

$$\pi_i(j) = \frac{a_i(j)}{1 + \sum_{s>0} a_i(s)}, \ \ a_i(s) = \prod_{l=1}^{s} \frac{\bar{\lambda}_i}{\min(l, \ r_i)\mu_i}, \ \ s > 0, \ a_i(0) = 1, \ \ j \geq 0. \tag{3.2}$$

Describe the network J by Markov process $z^J(t) = (n_1^J(t), \ldots, n_m^J(t))$ where $n_i^J(t)$ is a number of customers in i-th serving center at the moment t. If the formulas (3.1), (3.2) are true for $i = 1, 2, ..., m$, then Markov process $z^J(t)$ is ergodic [2], [7] and its stationary distribution P may be calculated by the formula [5]:

$$P(n_1, ..., n_m) = \prod_{i=1}^{m} \pi_i(n_i), \ \ n_1 \geq 0, ..., n_m \geq 0. \tag{3.3}$$

Put on branches (i, j), $1 \leq i, j \leq m$, of the network J queuing systems with infinite number of servers, which have exponential distributions of serving times with the parameters μ_{ij} (switching centers). A role of switching centers is only a service of customers without a redefinition of their routes. Denote such a network by J_1 and describe it by Markov process $z_1^J(t) = (n_i^J(t), \ n_{ij}^J(t), \ 1 \leq i, \ j \leq m)$, where $n_{ij}^J(t)$ is a number of customers in switching center arranged at (i, j) branch at the moment t. If

$$\pi_{ij}(n_{ij}) = \frac{\exp(-a_{ij})}{n_{ij}!}(a_{ij})^{n_{ij}}, \ \ a_{ij} = \frac{\bar{\lambda}_i \theta_{ij}}{\mu_{ij}} \tag{3.4}$$

then the formula (3.3) for the process $z_1^J(t)$ may be rewritten [5] as follows

$$P_1(n_i, \ n_{ij}, \ 1 \leq i, \ j \leq m) = P(n_1, ..., n_m) \left[\prod_{1 \leq i, j \leq m} \pi_{ij}(n_{ij}) \right], \tag{3.5}$$

$$n_i \geq 0,\ n_{ij} \geq 0,\ 0 \leq i, j \leq m.$$

Divide now a set of branches (i, j), $1 \leq i, j \leq m$, into nonintersecting subsets $K_1, \ldots, K_r$ and denote

$$A_{K_s} = \sum_{(ij)\in K_s} a_{ij},\ 1 \leq s \leq r.$$

Unite switching centers arranging on branches $(i, j) \in K_s$, $1 \leq s \leq r$, into common switching centers K_s with infinite number of servers and serving intensities depending on a route of each customer. Denote by J_2 a queuing network with so defined switching centers.

Introduce random process $z_2^J(t) = (n_i^J(t),\ 1 \leq i \leq m,\ N_s^J(t),\ 1 \leq s \leq r)$ describing the network J_2. Here $N_s^J(t)$ is a number of customers in the switching center K_s at the moment t. It is obvious that

$$N_s^J(t) = \sum_{(ij)\in K_s} n_{ij}^J(t),\ 1 \leq s \leq r, \tag{3.6}$$

and the process $z_2^J(t)$ is not Markov one.

Theorem 8. *The process $z_2^J(t)$ is ergodic and its limit distribution P_2 is following*

$$P_2(n_i,\ 1 \leq i \leq m,\ N_s,\ 1 \leq s \leq r) = P(n_1, \ldots, n_m)\left[\prod_{1\leq s\leq r} \psi_s(N_s)\right], \tag{3.7}$$

$$n_i \geq 0,\ N_s \geq 0,\ \psi_s(N_s) = \frac{\exp(-A_{K_s})}{N_s!}(A_{K_s})^{N_s},\ 1 \leq i \leq m,\ 1 \leq s \leq r.$$

The proof of the theorem 8 is based on ergodicity theorems [2], [7] and on the formulas (3.4) - (3.6) and on the following property of Poisson distribution: if independent random variables x, y have Poisson distributions with parameters a, b, then their sum $x+y$ has Poisson distribution with the parameter $a+b$.

3.2. Closed Networks

Consider closed network G (partial case of BCMP queuing network). Suppose that it has m multi server centers and N customers and $i-$ th center has $r_i \leq \infty$ servers (each server has exponentially distributed serving times with intensity μ_i) and allows infinite queue, $i = 1, 2, \ldots, m$. Determine a customer route in the network G by the route matrix $\Theta_G = (\theta_{ij})_{i,j=1}^m$, where θ_{ij} is a probability to transit from $i-$th serving center to $j-$th serving center and suppose that the route matrix is indivisible:

$$\forall i, j \in \{1, \ldots, m\}\ \exists\ i_1,\ i_2, \ldots i_r \in \{1, \ldots, m\}\ :\ \theta_{ii_1} > 0,\ \theta_{i_1 i_2} > 0, \ldots, \theta_{i_r j} > 0.$$

So for any $B > 0$ there is a single solution $(\overline{\lambda}_1, \overline{\lambda}_2, \ldots \overline{\lambda}_m)$ of the system

$$\sum_{i=1}^m \overline{\lambda}_i = B,\ (\overline{\lambda}_1, \overline{\lambda}_2, \ldots, \overline{\lambda}_m) = (\overline{\lambda}_1, \overline{\lambda}_2, \ldots, \overline{\lambda}_m)\Theta_G.$$

Analogously to previous subsection describe the network G by Markov process $z^G(t) = (n_1^G(t), \ldots, n_m^G(t))$. Suppose that for some $B > 0$ the formula (3.1) is true and (3.2) takes place. Then Markov process $z^G(t)$ is ergodic [2], [7] and its stationary distribution Q may be calculated by the formula [5]

$$Q(n_1,...,n_m) = D^{-1}P(n_1,...,n_m), \; n_1 \geq 0,...,n_m \geq 0, \; \sum_{i=1}^{m} n_i = N, \tag{3.8}$$

where

$$D = \sum_{n_1 \geq 0,...,n_m \geq 0, \, n_1+...+n_m=N} P(n_1,...,n_m).$$

Like in previous subsection put on branches (i,j), $1 \leq i,j \leq m$, of the network G queuing systems with infinite number of servers which have exponential distributions of serving times with the parameters μ_{ij} (switching centers). Denote such a network by G_1 and describe it by Markov process $z_1^G(t) = (n_i^G(t), \; n_{ij}^G(t), \; 1 \leq i, \; j \leq m)$. The formula (3.8) for the process $z_1^G(t)$ may be rewritten [5] as follows

$$Q_1(n_i, \; n_{ij}, \; 1 \leq i, \; j \leq m) = D_1^{-1}P_1(n_i, \; n_{ij}, \; 1 \leq i, \; j \leq m), \tag{3.9}$$

$$n_i \geq 0, \; n_{ij} \geq 0, \; 0 \leq i,j \leq m, \; \sum_{i=1}^{m} n_i + \sum_{i,j=1}^{m} n_{ij} = N,$$

$$D_1 = \sum_{n_i \geq 0, \; n_{ij} \geq 0, \; 0 \leq i,j \leq m, \; \sum_{i=1}^{m} n_i + \sum_{i,j=1}^{m} n_{ij} = N} P_1(n_i, \; n_{ij}, \; 1 \leq i, \; j \leq m).$$

Analogously to previous subsection divide now a set of branches (i,j), $1 \leq i,j \leq m$, into nonintersecting subsets $K_1,\ldots,K_r$. Unite switching centers arranging on branches $(i,j) \in K_s$, $1 \leq s \leq r$, into common switching centers K_s with infinite number of servers and serving intensities depending on a route of each customer. Denote by G_2 a queuing network with so defined switching centers. Describe the network G_2 by random process $z_2^G(t) = (n_i^G(t), \; 1 \leq i \leq m, \; N_s^G(t), \; 1 \leq s \leq r)$. Here $N_s^G(t)$ is a number of customers at the switching center K_s at the moment t,

$$N_s^G(t) = \sum_{(ij) \in K_s} n_{ij}^G(t), \; 1 \leq s \leq r, \tag{3.10}$$

Theorem 9. *The process $z_2^G(t)$ is ergodic and its limit distribution Q_2 is following*

$$Q_2(n_i, \; 1 \leq i \leq m, \; N_s, \; 1 \leq s \leq r) = D_2^{-1}P_2(n_i, \; 1 \leq i \leq m, \; N_s, \; 1 \leq s \leq r), \tag{3.11}$$

$$n_i \geq 0, \; N_s \geq 0, \; 1 \leq i \leq m, \; 1 \leq s \leq r, \; \sum_{i=1}^{m} n_i + \sum_{s=1}^{r} N_s = N,$$

$$D_2 = \sum_{n_i \geq 0, \; N_s \geq 0, \; 1 \leq i \leq m, \; 1 \leq s \leq r, \; \sum_{i=1}^{m} n_i + \sum_{s=1}^{r} N_s = N.} P_2(n_i, \; 1 \leq i \leq m, \; N_s, \; 1 \leq s \leq r).$$

The proof of the theorem 9 is analogous to the proof of the theorem 8 and is based on the formulas (3.4), (3.9), (3.10).

Remark 5. *The theorem 9 may be easily generalized onto the network G_2 in which r_i are replaced by* $\min(r_i,N)$, $i = 1,\ldots,m$, *and all switching centers contain not infinite but N servers and each center has maximal queue length equal not to infinity but to N.*

References

[1] Baum, D.; Tsitsiashvili, G.Sh. (2002) On Product Connection Theorems for Markov Chains. *International Journal on Pure and Applied Mathematics*, v. 1, number 2, 167 - 197.

[2] Borovkov, A.A. (1987) *A Limit Theorem for Queuing Networks. II. Theory Prob. Appl*, v. XXXII, number 2, 74-84.

[3] Jackson, J.R. (1957) *Networks of Waiting Lines. Oper. Res.*, v. 5, number 4, 518-521.

[4] Gabasov, R.; Kirillova, F.M. (1978) *Linear Programming Methods.* Minsk: BSU, 239 p. (in Russian).

[5] Gelenbe, E.; Munts R.R. (1976) *Probabilistic models of computer systems*. Acta Inform., part 1 (Exact results), v. 7 , number 1, 35-60.

[6] Gordon, K.D.; Newell, G.F. (1967) *Closed Queuing Systems with Exponential Servers*. Oper. Research, v. 15, number 2, 254-265.

[7] Ivchenko, G.I.; Kashtanov, V.A.; Kovalenko, I.N. (1982) Queuing Theory. Moscow: Vischaija Shkola. 256 p. (in Russian).

[8] Kelly, F.P. (1979) *Reversibility and Stochastic Networks*. John Wiley and Sons.

[9] Krilenko, A.V. (1997) Queuing Networks with Some Types of Customers and Immediate Serving and Rounds of Nodes by Customers. *Problems of Information Transmission*, v. 33, number 3, 91-101. (in Russian).

[10] Malinkovsky, Yu.V.; Nikitenko, O.A. (2000) Stationary Distributions of Networks States with Rounds and Negative Customers. *Authomatics and Telemechanics*,v. 8, 79-85. (in Russian).

[11] Pittel, B. (1979) Closed exponential networks of queues with saturation: the Jackson-type stationary distribution and its asymptotic analysis. *Math. Oper. Res.*,v. 4, 357-378.

[12] Tsitsiashvili, G.Sh.; Osipova, M.A.; Koliev, N.V.; Baum, D. (2002) A Product Theorem for Markov Chains with Applications to PF Queuing Networks. *Annals of Operations Research*, v. 113, 141-154.

[13] Serfozo, R.F. (2002) Reversible Markov Processes on General Spaces: Spatial Birth-Death and Queueing Processes. *Electonical Journal îInformation Processesî* ,v. 2, number 2, 259-261.

[14] Zhu, Y. (1994) Markovian Queueing Networks in Random Environment. *Operations Research Letters*, v. 15, 11–17.

In: Focus on Probability Theory
Editor: Louis R. Velle, pp. 109-125
ISBN 1-59454-474-3

Chapter 6

ASYMPTOTIC PROPERTIES OF A PICKANDS TYPE ESTIMATOR OF THE EXTREME VALUE INDEX

Laurent Gardes and Stéphane Girard
SMS/LMC-IMAG, Université Grenoble 1,
BP 53, 38041 Grenoble Cedex 9, France

Abstract

One of the main goals of extreme value analysis is to estimate the probability of rare events given a sample from an unknown distribution. The upper tail behavior of this distribution is described by the extreme value index. We present a new estimator of the extreme value index adapted to any domain of attraction. Its construction is similar to the one of Pickands' estimator. Its weak consistency as well as its asymptotic distribution are established and a bias reduction method is proposed.

1 Introduction

Suppose one is given a sequence $X_1, \ldots, X_n$ of independent and identically distributed (i.i.d.) observations from some distribution function F. Suppose there exist sequences $a_n > 0$ and b_n and some $\xi \in \mathbb{R}$ such that:

$$\lim_{n \to \infty} \mathbb{P}\left[\frac{\max(X_1, \ldots, X_n) - b_n}{a_n} \leq x\right] = G_\xi(x), \tag{1}$$

with $G_\xi(x) = \exp[-(1+\xi x)_+^{-1/\xi}]$ if $\xi \neq 0$ and $G_0(x) = \exp[-e^{-x}]$, where $y_+ = \max(0, y)$. Necessary and sufficient conditions on F for the convergence (1) to the extreme value distribution G_ξ can be found in [13]. The aim of this chapter is the definition of a new estimator of the extreme value index $\xi \in \mathbb{R}$. This parameter drives the decay of the tail distribution: as a power function if $\xi > 0$ (Pareto, Burr, Student's, Log-gamma distributions, etc ...), exponentially if $\xi = 0$ (Exponential, Normal, Log-normal, Gamma distributions, etc ...) and with finite right endpoint if $\xi < 0$ (Uniform, Beta, Reversed Pareto, Reversed Burr distributions, etc ...). The knowledge of ξ is for example of high interest for extreme quantile estimation which arises in a lot of applications [10] such as finance, insurance, hydrology, etc ... There is a substantial number of publications dedicated to the estimation of this extreme value index, especially on the heavy tailed distribution context ($\xi > 0$) (see Beirlant et *al.* [2], Feueverger and Hall [11] and, for a recent overview of this literature, see Csörgo and Viharos [5]). The most popular estimator in this case is the Hill estimator [15] defined by:

$$\hat{\xi}^H_{k,n} = \frac{1}{k}\sum_{i=1}^{k} \ln(X_{n-i+1,n}) - \ln(X_{n-k,n}), \text{ for } k = 1,\dots,n-1,$$

where $X_{1,n} \leq \dots \leq X_{n,n}$ correspond to the random variables $X_1,\dots,X_n$ rearranged in ascending order. The consistency and the asymptotic normality of this estimator are proved for example by Davis and Resnick [6], Csörgo and Mason [4], etc ...
The general case $\xi \in \mathbb{R}$ has been less extensively studied. Dekkers, Einmahl and de Haan [9] have adapted the estimator proposed by Hill to this situation. Another estimator was proposed by Pickands [16]:

$$\hat{\xi}^P_{k,n} = \frac{1}{\ln(2)} \ln\left(\frac{X_{n-k+1,n} - X_{n-2k+1,n}}{X_{n-2k+1,n} - X_{n-4k+1,n}}\right), \text{ for } k = 1,\dots,\lfloor n/4 \rfloor,$$

where $\lfloor x \rfloor$ denotes the integer part of x. Weak and strong consistency as well as asymptotic normality of $\hat{\xi}^P_{k,n}$ were established by Dekkers and de Haan [8]. Proofs are based on the following well known result: Let U be the tail quantile function of the distribution function F defined by

$$U(x) = \left(\frac{1}{1-F(x)}\right)^{\leftarrow}$$

(the arrow means inverse function) and let

$$\varphi_t(x) = \int_1^x u^{t-1} du, \ x > 0, \ t \in \mathbb{R}.$$

Relation (1) holds if and only if (see de Haan [7]), there exist a positive measurable function a such that uniformly locally on $x > 0$,

$$\lim_{t\to\infty} \frac{U(tx) - U(t)}{a(t)} = \varphi_\xi(x). \tag{2}$$

Clearly, relation (2) implies that uniformly locally on $x, y > 0, y \neq 1$,

$$\lim_{t\to\infty} \frac{U(tx) - U(t)}{U(ty) - U(t)} = \frac{\varphi_\xi(x)}{\varphi_\xi(y)}. \tag{3}$$

Thus, by substituting in (3) U by $\hat{U}_n = [1/(1-\hat{F}_n)]^{\leftarrow}$ ($\hat{F}_n$ denoting the empirical distribution function), t by $n/(2k)$, x by $1/2$ and y by 2, and remarking that $\hat{U}_n(n/k) = X_{n-k+1,n}$, we have asymptotically

$$2^{-\xi} \frac{X_{n-k+1,n} - X_{n-2k+1,n}}{X_{n-2k+1,n} - X_{n-4k+1,n}} = 1. \tag{4}$$

Pickands' estimator $\hat{\xi}^P_{k,n}$ is the solution of the equation (4). One can notice that this estimator does not take into account of the extreme observations $X_{n-k+2,n},\dots,X_{n,n}$.

In the next section, we define a new estimator of the extreme value index ξ when $\xi \in \mathbb{R}$. This estimator is similar to the one of Pickands but exploiting the information given by the spacing between $X_{n-k+1,n}$ and $X_{n,n}$. Weak consistency and asymptotic distribution are established in section 3 and a bias corrected estimator is introduced. Section 4 is devoted to the proofs of the main results.

2 Estimation of the Extreme Value Index

We propose to estimate the extreme value index $\xi \in \mathbb{R}$ by $\hat{\xi}_{k,n}$ defined as the root of the equation in θ:

$$\left\{\frac{\varphi_\theta(1/k')}{\varphi_\theta(1/k)}\right\}\frac{X_{n-k+1,n}-X_{n,n}}{X_{n-k'+1,n}-X_{n,n}} = 1, \text{ for } 1 < k' < k < n. \tag{5}$$

Existence of the estimator $\hat{\xi}_{k,n}$ and unicity are ensured by the following result.

Lemma 1 If $1 < k' < k$, then (5) admits an unique solution.

This estimator applies to all real ξ and, as Pickands' estimator, remains unaffected when the scale or location of the data are changed. One can justify the definition of $\hat{\xi}_{k,n}$ by the following two lemmas:

Lemma 2 Suppose that relation (1) holds. Then, if $\xi < 0$,

$$\lim_{t\to\infty,\ x\to\infty} \frac{U(tx)-U(t)}{a(t)} = -\frac{1}{\xi}$$

and if $\xi \geq 0$,

$$\lim_{t\to\infty,\ x\to\infty} \frac{U(tx)-U(t)}{a(t)} = +\infty.$$

As a consequence of Lemma 2, we have:

Lemma 3 Suppose that relation (1) holds. Then,

$$\left\{\frac{\varphi_\xi(y)}{\varphi_\xi(x)}\right\}\frac{U(tx)-U(t)}{U(ty)-U(t)} \to 1, \tag{6}$$

as $t \to \infty$, $x \to 0$ with $ty \to \infty$ and $x/y \to d > 0$.

Lemma 2 and Lemma 3 can be seen as an extension of respectively (2) and (3) when x and y are going to zero or infinity. The proofs of Lemma 2 and Lemma 3 are postponed to the Appendix. By substituting in (6) U by $\hat{U}_n$, t by n, x by $1/k'(n) = 1/k'$ and y by $1/k(n) = 1/k$ with $k/k' \to c > 1$, $k \to \infty$ and $k/n \to 0$ as $n \to \infty$, we have asymptotically:

$$\left\{\frac{\varphi_\xi(1/k')}{\varphi_\xi(1/k)}\right\}\frac{X_{n-k+1,n}-X_{n,n}}{X_{n-k'+1,n}-X_{n,n}} = 1,$$

which is an intuitive justification for the definition of $\hat{\xi}_{k,n}$. The next section is dedicated to the study of $\hat{\xi}_{k,n}$ asymptotical properties.

3 Main Results

3.1 Asymptotic Properties

We first state the weak consistency of $\hat{\xi}_{k,n}$ under some conditions on k and k'.

Theorem 1 Suppose that relation (1) holds. If $k/k' \longrightarrow c > 1$, $k \longrightarrow \infty$ and $k/n \longrightarrow 0$ as $n \longrightarrow \infty$, then $\hat{\xi}_{k,n} \xrightarrow{\text{P}} \xi$.

Remark 1 Similar conditions on k are used by Dekkers and de Haan [8] to prove that Pickands' estimator $\hat{\xi}^P_{k,n}$ is weakly consistent.

To establish the asymptotic distribution of the estimator $\hat{\xi}_{k,n}$, additionnary conditions are introduced. The first of them is a cornerstone in all proofs of asymptotic normality for extreme value estimators.

$(H1)$ – U has a positive derivative and there exist a slowly varying function ℓ such that $U'(x) = x^{\xi-1}\ell(x)$.

We refer to [3] for more details on slow variation theory. The next condition controls the uniform rate of convergence of $\ell(tx)/\ell(x)$ to 1 as $x \longrightarrow \infty$. Let $\delta = \min(-\xi, 1/2)$ and introduce the random variables $K_{k,n} = \bar{F}(X_{n-k+1,n})/\bar{F}(X_{n,n})$ and $N_n = 1/\bar{F}(X_{n,n})$ where $\bar{F}$ is the survival function ($\bar{F} = 1 - F$).

$(H2)$ –

$$\varphi_\delta(k') \sup_{t \in [1, K_{k',n}]} \left| \frac{\ell(tN_n/K_{k',n})}{\ell(N_n/K_{k',n})} - 1 \right| \xrightarrow{\text{P}} 0.$$

Our second main result is the following:

Theorem 2 Let $V_k(\xi) = \varphi_\delta(k)[(\ln(k)-1)\mathbb{I}\{\xi \geq 0\} + 1]$. Under the conditions of Theorem 1 (with $k = ck'$) and if $(H1)$ and $(H2)$ are satisfied, we have for all $t \in \mathbb{R}$:

$$\lim_{n\to\infty} \mathbb{P}[V_k(\xi)(\hat{\xi}_{k,n} - \xi) \leq t] = \begin{cases} \exp(-e^{-t}) & \text{if } 0 < \xi, \\ \exp(-e^{-t/2}) & \text{if } \xi = 0, \\ \exp\left[-\left[1 + t\ln(c)/\varphi_\xi(1/c)\right]^{-1/\xi}\right] & \text{if } -1/2 < \xi < 0, \\ \Phi\left[-tc^{-\xi}\ln(c)/(2\xi\sigma)\right] & \text{if } \xi < -1/2, \end{cases} \tag{7}$$

where $\sigma = c^{-\xi}(c-1)^{1/2}$ and Φ is the cumulative distribution function of the standard normal distribution.

Remark 2

i) Theorem 2 states that the asymptotic distribution of $\hat{\xi}_{k,n}$ is Gaussian if $\xi < -1/2$ and an extreme value distribution if $\xi > -1/2$. If $\xi = -1/2$, we prove that $V_k(\xi)(\hat{\xi}_{k,n} - \xi)$ converges to a non-degenerate distribution with non explicit cumulative distribution function. In fact, as it will appear in the next section, the limit distribution of $\hat{\xi}_{k,n}$ is driven by $k'/K_{k',n}$ if $\xi < -1/2$, by Y_n if $\xi > -1/2$ with

$$Y_n = \sqrt{k}/\sigma[\varphi_\xi(K_{k',n}/K_{k,n}) - \varphi_\xi(1/c)]$$

and by both of them if $\xi = -1/2$.

ii) $(H1)$ and $(H2)$ are second order conditions on the tail quantile function U. We refer to [12] for a possible simplification of these condictions. Similar ones are used by Dekkers and de Haan [8] to establish the asymptotic distribution of Pickands' estimator.

iii) Let $\mu(\xi) = \gamma\mathbb{I}\{\xi > 0\} - [1 - \Gamma(1-\xi)]\varphi_\xi(1/c)/\ln(c)\mathbb{I}\{-1/2 < \xi < 0\}$, where γ is the Euler constant and Γ is the gamma function. Theorem 2 entails that $V_k(\xi)(\hat{\xi}_{k,n} - \xi)$ converges to a distribution of mean $\mu(\xi)$ if $\xi \neq 0$ and $\xi \neq -1/2$. This suggests to define the bias corrected estimator:

$$\hat{\xi}^*_{k,n} = \hat{\xi}_{k,n} - \frac{\mu(\hat{\xi}_{k,n})}{V_k(\hat{\xi}_{k,n})}.$$

3.2 Examples

Let $\alpha, \beta > 0$, $\theta \in \mathbb{R}\backslash\{0\}$ and define $\ln_2(x) = \ln(\ln(x))$, $x > 1$. The two following models of slowly varying functions ℓ are considered:

$$\ell(x) = \alpha + \theta x^{-\beta} + o(x^{-\beta}), \quad \textbf{(Model A)},$$

$$\ell(x) = \theta[\ln(x)]^{-\beta}\left\{1 + O\left(\frac{\ln_2(x)}{\ln(x)}\right)\right\}, \quad \textbf{(Model B)}.$$

Model A has been first introduced by Hall [14]. In both models, the parameter β tunes the decay of the slowly varying function ℓ. The conditions that should be satisfied by models A and B to insure convergence (7) are given in Corollary 1. In both cases, the best rate of convergence of $\hat{\xi}_{k,n}$ is also established. Some examples of distributions satisfying the assumptions of Corollary 1 are presented in Table 3.2. In the sequel, the following notation is adopted. Let (u_n) and (v_n) be two non negative deterministic sequences. The notation $u_n \asymp v_n$ means that

$$0 < \liminf \frac{u_n}{v_n} \leq \limsup \frac{u_n}{v_n} < \infty.$$

Corollary 1 Suppose that $k = ck'$, $k \to \infty$ and that F satisfies assumption $(H1)$ with a slowly varying function asymptotically monotone.

Table 1: Examples of distributions satisfying the assumptions of Corollary 1

Distribution	Cumulative distribution function	Model	β	Best rate of convergence
Weibull$_M$ ($\xi \in \mathbb{R}$)	$\exp[-(1+\xi x)^{-1/\xi}]$, for x such that $1+\xi x > 0$.	A	1	$\ln(n)$ if $0 < \xi$ $\ln^2(n)$ if $\xi = 0$ $n^{\delta/(1+\delta/\beta)-\varepsilon}$ if $\xi < 0$
Burr ($\xi > 0$)	$1-[w/(w+x^\tau)]^\lambda$, for $x > 0$, with w, λ, $\tau > 0$, $\xi = 1/(\lambda\tau)$,	A	$1/\lambda$	$\ln(n)$
Fréchet ($\xi > 0$)	$\exp(-x^{-1/\xi})$ for $x > 0$,	A	1	$\ln(n)$
Weibull ($\xi = 0$)	$1-\exp(-\lambda x^\tau)$, for $x > 0$, with λ, $\tau > 0$.	B	$1-1/\tau$	$\ln_2^2(n)$
Normal ($\xi = 0$)	$\int_{-\infty}^{x} 1/\sqrt{2\pi} e^{-t^2/2} dt$	B	1/2	$\ln_2^2(n)$.
Reversed Burr ($\xi < 0$)	$1-[w/(w+(x_F-x)^{-\tau})]^\lambda$, for $x < x_F$, with $\xi = 1/(\lambda\tau)$. w, λ, $\tau > 0$, $x_F \in \mathbb{R}$.	A	$1/\lambda$	$n^{\delta/(1+\delta/\beta)-\varepsilon}$

i) If ℓ belongs to Model A and if $\varphi_\delta(k')\,(n/k')^{-\beta} \to 0$ then convergence (7) holds. In this case, the best rate of convergence of $\hat{\xi}_{k,n}$ is given by:

$$V_k(\xi) \asymp \begin{cases} \ln(n) & \text{if } 0 < \xi, \\ \ln^2(n) & \text{if } \xi = 0, \\ n^{\delta/[1+\delta/\beta]-\varepsilon} & \text{if } \xi < 0, \end{cases}$$

where $\varepsilon \in]0, \delta/(1+\delta/\beta)[$ is arbitrarily small.

ii) If ℓ belongs to Model B and if $\varphi_\delta(k')\ln(k')/\ln(n) \to 0$ and $\varphi_\delta(k')\ln_2(n)/\ln(n) \to 0$ then, convergence (7) holds. Furthermore, the best rate of convergence of $\hat{\xi}_{k,n}$ is given by:

$$V_k(\xi) \asymp \begin{cases} \ln_2(n) & \text{if } 0 < \xi, \\ \ln_2^2(n) & \text{if } \xi = 0, \\ \ln^{1-\varepsilon}(n) & \text{if } \xi < 0, \end{cases}$$

where $\varepsilon \in]0, 1[$ is arbitrarily small.

Remark 3 This corollary points out the fact that the case $\xi < 0$ is more favorable to our estimator, i.e. its convergence is faster than in the case $\xi \geq 0$.

4 Proofs of the Main Results

This section is devoted to the proof of Theorem 1 and Theorem 2. Proofs of lemmas are postponed to the Appendix.

4.1 Preliminary Results

The following function will play an important role. Let

$$H_n(x) = \left\{ \frac{\varphi_x(1/k')}{\varphi_x(1/k)} \right\} \frac{X_{n-k+1,n} - X_{n,n}}{X_{n-k'+1,n} - X_{n,n}} = \left\{ \frac{\varphi_x(1/k')}{\varphi_x(1/k)} \right\} (1+Z_n),$$

with $Z_n = (X_{n-k+1,n} - X_{n-k'+1,n})/(X_{n-k'+1,n} - X_{n,n})$ and $x \in \mathbb{R}$.

Lemma 4 Under the conditions of Theorem 1,

i) $Z_n \stackrel{d}{=} [U(N_n/K_{k,n}) - U(N_n/K_{k',n})]/[U(N_n/K_{k',n}) - U(N_n)]$.

ii) $N_n \xrightarrow{\text{a.s.}} +\infty$, $K_{k,n} \xrightarrow{\text{a.s.}} +\infty$, $K_{k,n}/K_{k',n} \xrightarrow{\text{a.s.}} c$, $K_{k,n}/N_n \xrightarrow{\text{a.s.}} 0$ and $Z_n \xrightarrow{\text{a.s.}} \max(0, c^{-\xi} - 1)$ as $n \to \infty$.

iii) $k'/K_{k',n} \xrightarrow{d} \text{Exp}(1)$. If moreover $k = ck'$, then $Y_n \xrightarrow{d} \mathcal{N}(0,1)$.

4.2 Proof of Theorem 1

We shall need the following result:

Lemma 5 Suppose that relation (1) holds. If $\xi > 0$, for all $\eta_1 \in]0, \xi[$, $\eta_2 > 0$, there exist t_0, $\beta_1, \beta_2, \tilde{\beta_1}, \tilde{\beta_2} > 0$ such that, for all $t \geq t_0$ and $x > 0$,

$$\beta_1 x^{\xi - \eta_1} - \tilde{\beta_1} \leq \frac{U(tx) - U(t)}{a(t)} \leq \beta_2 x^{\xi + \eta_2} - \tilde{\beta_2}.$$

If $\xi = 0$, for all $\eta > 0$, there exist t_0, $\beta > 0$ such that, for all $t \geq t_0$ and $x > 0$,

$$\frac{U(tx) - U(t)}{a(t)} \leq \beta x^{\eta}.$$

Proof of Theorem 1 – We have to show that, for all $\varepsilon > 0$,

$$\lim_{n \to \infty} \mathbb{P}[|\hat{\xi}_{k,n} - \xi| > \varepsilon] = 0. \tag{8}$$

Remark that if $\xi \neq 0$, proving (8) for all $\varepsilon > 0$ reduces to demonstrate (8) for all $0 < \varepsilon < |\xi|$. Since H_n is a non-decreasing function (see proof of Lemma 1) and since $H_n(\hat{\xi}_{k,n}) = 1$, we have:

$$\mathbb{P}[|\hat{\xi}_{k,n} - \xi| > \varepsilon] = \mathbb{P}[H_n(\xi + \varepsilon) < 1] + \mathbb{P}[H_n(\xi - \varepsilon) > 1].$$

Consequently, to prove Theorem 1 it is sufficient to establish that $\mathbb{P}[H_n(\xi + \varepsilon) \geq 1] \to 1$ and $\mathbb{P}[H_n(\xi - \varepsilon) \leq 1] \to 1$ as $n \to \infty$. The following two expansions hold:

$$\frac{\varphi_t(1/k')}{\varphi_t(1/k)} = 1 + k^{-t} - k'^{-t} + o(k^{-t}) \text{ if } t > 0 \text{ and } \frac{\varphi_t(1/k')}{\varphi_t(1/k)} \to c^t \text{ as } n \to \infty \text{ if } t < 0. \tag{9}$$

The following two cases are considered separately:
If $\xi \geq 0$, (9) and Lemma 4 ii) imply that $H_n(\xi+\varepsilon) = 1 + Z_n + k^{-(\xi+\varepsilon)} - k'^{-(\xi+\varepsilon)} + o_{\mathrm{P}}[k^{-(\xi+\varepsilon)}]$ since $\xi+\varepsilon > 0$. Furthermore, from Lemma 4 i),

$$Z_n \stackrel{d}{=} \underbrace{\frac{U(N_n/K_{k,n}) - U(N_n/K_{k',n})}{a(N_n/K_{k',n})}}_{Z_{1,n}} \times \underbrace{\frac{a(N_n/K_{k',n})}{U(N_n/K_{k',n}) - U(N_n)}}_{Z_{2,n}}, \tag{10}$$

with

$$Z_{1,n} \xrightarrow{\text{a.s.}} \varphi_\xi(c^{-1}), \tag{11}$$

from (2) and Lemma 4 ii). Following Lemma 5, we have

$$\beta_1 K_{k',n}^{\xi-\eta_1} - \tilde{\beta_1} \leq -\frac{1}{Z_{2,n}} \leq \beta_2 K_{k',n}^{\xi+\eta_2} - \tilde{\beta_2}, \tag{12}$$

Since, from Lemma 4 iii), $k'/K_{k',n}$ converges to a standard exponential distribution, we deduce from (10)-(12) that $k'^{-(\xi+\varepsilon)}/Z_n \xrightarrow{\mathrm{P}} 0$ which entails that $H_n(\xi+\varepsilon) - 1 \stackrel{\mathrm{P}}{\sim} Z_n > 0$ i.e. that $\mathbb{P}[H_n(\xi+\varepsilon) \geq 1] \to 1$. Similarly, we prove that $\mathbb{P}[H_n(\xi-\varepsilon) \leq 1] \to 1$.
If $\xi < 0$, expansions (9) and Lemma 4 ii) imply that $H_n(\xi+\varepsilon) \to c^\varepsilon > 1$ since $\xi+\varepsilon > 0$. Thus, $\mathbb{P}[H_n(\xi+\varepsilon) \geq 1] \to 1$. In the same way, we prove that $\mathbb{P}[H_n(\xi-\varepsilon) \leq 1] \to 1$. ■

4.3 Proof of Theorem 2

Let us define the function:

$$\varphi_t^*(x) = \begin{cases} 1+tx & \text{if } t \neq 0, \\ e^x & \text{if } t = 0. \end{cases}$$

To prove Theorem 2, two auxiliary results are necessary. Lemma 6 is dedicated to the study of the function φ_t^*.

Lemma 6

I) For $x \in (0,\infty)$, $\varphi_t^*[\varphi_t(x)] = x^{t+\mathbb{I}\{t=0\}}$.

Let (u_n) and (v_n) be two sequences such that $u_n \sim v_n$ (i.e. $u_n/v_n \to 1$).

II) Let $t=0$. If $u_n \to \infty$ and $u_n - v_n \to \alpha$ then $\varphi_t^*(v_n) \sim \varphi_t^*(u_n)e^{-\alpha}$.

III) Let $t \neq 0$. If $u_n \to \infty$ then $\varphi_t^*(v_n) \sim \varphi_t^*(u_n)$.

IV) Let $t \neq 0$. If $u_n \to -1/t$ with $v_n = u_n(1+\varepsilon_n)$, then:

i) If moreover $\varepsilon_n/\varphi_t^*(u_n) \sim \alpha_n$ where α_n does not converge to ∞ or to 1, then $\varphi_t^*(v_n) \sim \varphi_t^*(u_n)(1-\alpha_n)$.

ii) If moreover $\varepsilon_n/\varphi_t^*(u_n) \to \infty$ then $\varphi_t^*(v_n) \sim -\varepsilon_n$.

The proof of this basic result is not detailed here. Clearly, the distribution of $H_n(x)$ is determined by Z_n. The following lemma provides the asymptotic distribution of Z_n.

Lemma 7 Under the conditions of Theorem 2,

$$\lim_{n\to\infty} \mathbb{P}\left[k'^{\delta-\mathbb{I}\{\xi=0\}}\varphi_\xi^*\left(-\frac{\varphi_\xi(1/c)}{Z_n}\right)\le t\right] = \begin{cases} \exp(-t^{-1/\xi}) & \text{if } 0<\xi, \\ \exp(-t^{-1}) & \text{if } \xi=0, \\ 1-\exp(-t^{-1/\xi}) & \text{if } -1/2<\xi<0, \\ \mathbb{P}[T<t\sqrt{c}] & \text{if } \xi=-1/2, \\ \Phi\left[-t\varphi_\xi(1/c)\sqrt{c}/\sigma\right] & \text{if } \xi<-1/2, \end{cases}$$

where the random variable T is defined as the limit in distribution of

$$T_n = \sqrt{\frac{k}{K_{k',n}}} + \frac{\sigma}{\varphi_\xi(1/c)}Y_n,$$

which is non-degenerate from Lemma 4 iii).

Proof of Theorem 2 – Let $F_n(t) = \mathbb{P}[V_k(\xi)(\hat{\xi}_{k,n}-\xi)\le t]$. We have,

$$F_n(t) = \mathbb{P}\left[\hat{\xi}_{k,n}\le \xi + t/V_k(\xi)\right] = \mathbb{P}[H_n(\xi+t/V_k(\xi))\ge 1] = \mathbb{P}\left[(1+Z_n)\frac{\varphi_{-\xi-t/V_k(\xi)}(k/c)}{\varphi_{-\xi-t/V_k(\xi)}(k)}\ge 1\right],$$

since H_n is a non-decreasing function and since $H_n(\hat{\xi}_{k,n}) = 1$. Routine calculations yield:

$$F_n(t) = \mathbb{P}\left[-\frac{\varphi_\xi(1/c)}{Z_n}\le t_n\right],$$

with

$$t_n = -\varphi_\xi(1/c)\frac{\varphi_{-\xi-t/V_k(\xi)}(k/c)}{\varphi_{-\xi-t/V_k(\xi)}(k) - \varphi_{-\xi-t/V_k(\xi)}(k/c)}.$$

Remarking that φ_ξ^* is an increasing function for $\xi\ge 0$ and decreasing for $\xi<0$, we have,

$$F_n(t) = \begin{cases} \mathbb{P}\left[(k')^{\delta-\mathbb{I}\{\xi=0\}}\varphi_\xi^*\left(-\frac{\varphi_\xi(1/c)}{Z_n}\right)\le (k')^{\delta-\mathbb{I}\{\xi=0\}}\varphi_\xi^*(t_n)\right] & \text{if } \xi\ge 0, \\ \mathbb{P}\left[(k')^{\delta}\varphi_\xi^*\left(-\frac{\varphi_\xi(1/c)}{Z_n}\right)\ge (k')^{\delta}\varphi_\xi^*(t_n)\right] & \text{if } \xi<0. \end{cases} \tag{13}$$

The asymptotic behavior of the left hand side random term

$$(k')^{\delta-\mathbb{I}\{\xi=0\}}\varphi_\xi^*\left(-\frac{\varphi_\xi(1/c)}{Z_n}\right)$$

is given by Lemma 7. Let us now focus on the right hand side term $(k')^{\delta-\mathbb{I}\{\xi=0\}}\varphi_\xi^*(t_n)$. Different cases have to be considered:

If $\xi>0$, the following sequence of asymptotic equivalences holds

$$t_n \sim k^{\xi+t/V_k(\xi)}\frac{\varphi_\xi(1/c)}{1-c^{\xi+t/V_k(\xi)}} \sim k^{\xi+t/V_k(\xi)}c^{-\xi}/\xi.$$

Since $V_k(\xi) \sim \ln(k)/\xi$, we have that $k^{t/V_k(\xi)} \longrightarrow e^{t\xi}$ as $n \longrightarrow \infty$. Thus, $t_n \sim (k')^{\xi}/\xi e^{t\xi} \longrightarrow \infty$ as $n \longrightarrow \infty$. Lemma 6 III) implies that $\varphi_\xi^*(t_n) \sim 1 + (k')^{\xi} e^{t\xi} \sim (k')^{\xi} e^{t\xi}$. Thus,

$$\lim_{n\to\infty} (k')^{\delta - \mathbb{I}\{\xi=0\}} \varphi_\xi^*(t_n) = e^{t\xi}. \tag{14}$$

If $\xi = 0$, we have

$$t_n = \ln(c) \frac{c^{t/\ln^2(k)} - k^{t/\ln^2(k)}}{1 - c^{t/\ln^2(k)}}.$$

Using the expansions,

$$c^{t/\ln^2(k)} = 1 + t\frac{\ln(c)}{\ln^2(k)} + O\left(\frac{1}{\ln^4(k)}\right) \text{ and } k^{t/\ln^2(k)} = 1 + t\frac{1}{\ln(k)} + t^2\frac{1}{2\ln^2(k)} + o\left(\frac{1}{\ln^2(k)}\right),$$

we find that

$$t_n = \ln(k)\left[1 + \frac{t/2 - \ln(c)}{\ln(k)} + o\left(\frac{1}{\ln(k)}\right)\right].$$

Since $t_n \longrightarrow \infty$ and $\ln(k) - t_n \longrightarrow -t/2 + \ln(c)$ as $n \longrightarrow \infty$, Lemma 6 II) implies that

$$\varphi_\xi^*(t_n) \sim \varphi_\xi^*(\ln(k)) \exp[t/2 - \ln(c)] = k' \exp(t/2).$$

Thus,

$$\lim_{n\to\infty} (k')^{\delta - \mathbb{I}\{\xi=0\}} \varphi_\xi^*(t_n) = e^{t/2}. \tag{15}$$

When $\xi < 0$, we have:

$$t_n = -\frac{1}{\xi}\frac{c^{-\xi} - 1}{c^{-\xi} - c^{t/V_k(\xi)}} c^{t/V_k(\xi)} [1 - (k')^{\xi + t/V_k(\xi)}].$$

Remarking that

$$c^{t/V_k(\xi)} = 1 + t\frac{\ln(c)}{V_k(\xi)} + O\left(\frac{1}{V_k(\xi)}\right),$$

and that $(k')^{t/V_k(\xi)} = 1 + o(1)$ lead to the following expansion:

$$\begin{aligned} t_n &= -\frac{1}{\xi}\left[1 + t\frac{\ln(c)}{V_k(\xi)}\right]\left[1 + o\left(\frac{1}{V_k(\xi)}\right)\right]\left[1 + t\frac{\ln(c)}{(c^{-\xi} - 1)V_k(\xi)} + o\left(\frac{1}{V_k(\xi)}\right)\right] \\ &\times \left[1 - (k')^{\xi} + o(k^{\xi})\right] = u_n(1 + \varepsilon_n), \end{aligned}$$

with $u_n = -1/\xi[1 + t\ln(c)/V_k(\xi)]$ and

$$\varepsilon_n = t\frac{\delta \ln(c)}{c^{-\xi} - 1} k^{-\delta} - (k')^{\xi} + o(k^{-\delta}) + o(k^{\xi}).$$

Two situations have to be considered:

If $-1/2 \le \xi < 0$, it follows that

$$\lim_{n\to\infty} \frac{\varepsilon_n}{\varphi_\xi^*(u_n)} = -\frac{1}{t\xi\ln(c)}\left[\frac{t\xi\ln(c)}{c^{-\xi} - 1} + c^{-\xi}\right],$$

and Lemma 6 IV) i) implies

$$\varphi_\xi^*(t_n) \sim \varphi_\xi^*(u_n)\left[1+\frac{1}{t\xi\ln(c)}\left(\frac{t\xi\ln(c)}{c^{-\xi}-1}+c^{-\xi}\right)\right]$$

and thus

$$\lim_{n\to\infty}(k')^{\delta-\mathbb{I}\{\xi=0\}}\varphi_\xi^*(t_n) = 1+t\frac{\xi\ln(c)}{c^{-\xi}-1}. \tag{16}$$

If $\xi < -1/2$, $\varepsilon_n/\varphi_\xi^*(u_n) \to (1-c^{-\xi})^{-1}$ as $n\to\infty$. Thus, Lemma 6 IV) i) yields

$$\lim_{n\to\infty}(k')^{\delta-\mathbb{I}\{\xi=0\}}\varphi_\xi^*(t_n) = -\frac{t}{2}\ln(c)\frac{c^{-\xi-1/2}}{1-c^{-\xi}}. \tag{17}$$

Collecting (13)-(17) with Lemma 7 concludes the proof. ■

5 Appendix

Proof of Lemma 1 – For $x\in\mathbb{R}$, let

$$H_n(x) = \left\{\frac{\varphi_x(1/k')}{\varphi_x(1/k)}\right\}\frac{X_{n-k+1,n}-X_{n,n}}{X_{n-k'+1,n}-X_{n,n}}.$$

Remarking that H_n is a continuous function and that

$$\lim_{x\to-\infty} H_n(x) = 0 \text{ and } \lim_{x\to+\infty} H_n(x) > 1,$$

it is sufficient to prove that H_n is an increasing function or equivalently that the derivative of

$$G_n(x) = \frac{\varphi_x(1/k')}{\varphi_x(1/k)} = \frac{(k')^{-x}-1}{k^{-x}-1},\ x\neq 0,$$

is positive. For $x\neq 0$ we have:

$$\begin{aligned} G_n'(x) &= \frac{k^{-x}\ln(x)[(k')^{-x}-1]-(k')^{-x}\ln(k')(k^{-x}-1)}{(k^{-x}-1)^2} \\ &= \frac{(kk')^{-x}[\ln(k)-\ln(k')]+(k')^{-x}\ln(k')(k^{-x}-1)}{(k^{-x}-1)^2}. \end{aligned}$$

Introducing $\ln(k') = u$ and $\ln(k) = u+v$, $v>0$, yields

$$G_n'(x) = \frac{v\exp[-x(2u+v)]+u\exp[-ux]-(u+v)\exp[-x(u+v)]}{(k^{-x}-1)^2}.$$

Thus, to prove that $G_n'(x)>0$ for $x\neq 0$, it is sufficient to demonstrate that for $x\neq 0$:

$$\frac{v}{u+v}\exp[-x(2u+v)]+\frac{u}{u+v}\exp[-xu] > \exp[-x(u+v)]. \tag{18}$$

Let us introduce the following notations:

$$\frac{1}{p_1} = \frac{v}{u+v},\ \frac{1}{p_2} = \frac{u}{u+v},\ a = \exp\left[-x\frac{v(2u+v)}{u+v}\right] \text{ and } b = \exp\left[-x\frac{u^2}{u+v}\right].$$

Inequality (18) can be rewritten:

$$\frac{a^{p_1}}{p_1}+\frac{b^{p_2}}{p_2} > \exp[-x(u+v)].$$

Since $1/p_1+1/p_2=1$ and $a\neq b$, one can use the well known inequality:

$$\frac{a^{p_1}}{p_1}+\frac{b^{p_2}}{p_2} > ab,$$

and the conclusion of the proof is then straightforward after remarking that:

$$ab=\exp\left[-x\frac{v(2u+v)+u^2}{u+v}\right]=\exp[-x(u+v)].$$

■

Proof of Lemma 2 – First, we focus on the case $\xi<0$. Recall that one can take in relation (2) $a(t)=\xi[U(t)-x_F]$ where x_F is the right endpoint of the distribution function F. Thus,

$$\frac{U(tx)-U(t)}{a(t)}=-\frac{1}{\xi}-\frac{x_F-U(tx)}{\xi[U(t)-x_F]}<-\frac{1}{\xi}, \tag{19}$$

since $x_F>U(t)$ for all t. Furthermore, since U is a non-decreasing function, we have that for all $A>0$ and $x\geq A$,

$$\frac{U(tx)-U(t)}{a(t)}\geq\frac{U(tA)-U(t)}{a(t)}.$$

Using (2), we have that for all $A,\varepsilon>0$ there exist T such that for all $t\geq T$ and $x\geq A$,

$$\frac{U(tx)-U(t)}{a(t)}\geq-\frac{1}{\xi}(1+\varepsilon)(1-A^{\xi}). \tag{20}$$

Using (19) and (20), we conclude the proof for $\xi<0$. Second, suppose that $\xi=0$ (the proof for $\xi>0$ is quite similar). Using relation (2) we have that for all $A,\varepsilon>0$ there exist T such that for all $t\geq T$ and $x\geq A$,

$$\frac{U(tx)-U(t)}{a(t)}\geq(1+\varepsilon)\ln(A),$$

which concludes the proof. ■

Proof of Lemma 3 – Remark that

$$\frac{U(tx)-U(t)}{U(ty)-U(t)}=1+z(t,x,y),$$

with

$$z(t,x,y)=\frac{U(tx)-U(ty)}{U(ty)-U(t)}=\underbrace{\frac{U(tyx/y)-U(ty)}{a(ty)}}_{z_1(t,x,y)}\times\underbrace{\frac{a(ty)}{U(ty)-U(ty/y)}}_{z_2(t,x,y)}.$$

Since $y \longrightarrow \infty$ and $x/y \rightarrow d$, relation (2) implies that $z_1(t,x,y)$ converges to $\varphi_\xi(d)$. Since $y \longrightarrow \infty$ and $ty \longrightarrow \infty$, Lemma 2 implies that $z_2(t,x,y)$ converges to $\min(0,\xi)$. Consequently, $1+z(t,x,y) \longrightarrow 1$ if $\xi \geq 0$ and $1+z(t,x,y) \rightarrow c^\xi$ if $\xi < 0$. Remarking that $\varphi_\xi(y)/\varphi_\xi(x) \rightarrow 1$ if $\xi \geq 0$ and $\varphi_\xi(y)/\varphi_\xi(x) \rightarrow c^{-\xi}$ if $\xi < 0$ concludes the proof. ■

Proof of Lemma 4 – Let $W_1,\ldots,W_n$ be independent standard uniform random variables and $W_{1,n} \leq \ldots \leq W_{n,n}$ the corresponding order statistics.

i) Remarking that

$$(1-K_{i,n}/N_n)_{1\leq i\leq n} = (F(X_{n-i+1,n}))_{1\leq i\leq n} \stackrel{d}{=} (W_{n-i+1,n})_{1\leq i\leq n},$$

it follows that

$$\begin{aligned}(U(N_n/K_{i,n}))_{1\leq i\leq n} = (F^{\leftarrow}(1-K_{i,n}/N_n))_{1\leq i\leq n} &\stackrel{d}{=} (F^{\leftarrow}(W_{n-i+1,n}))_{1\leq i\leq n}\\ &\stackrel{d}{=} (X_{n-i+1,n})_{1\leq i\leq n},\end{aligned}$$

which conclude the demonstration.

ii) We have $K_{k,n} \stackrel{d}{=} W_{k,n}/W_{1,n}$. Using Rényi representation [1] p.72, $K_{k,n} \stackrel{d}{=} T_k/T_1$ where T_k is the sum of k independent standard exponential random variables. Another use of Rényi representation leads to $K_{k,n} \stackrel{d}{=} 1/W_{1,k-1}$. Since $W_{1,k-1} \stackrel{\text{a.s.}}{\longrightarrow} 0$ as $n \longrightarrow \infty$, we prove that $K_{k,n} \stackrel{\text{a.s.}}{\longrightarrow} \infty$ as $n \rightarrow \infty$. Similarly, $N_n \stackrel{\text{a.s.}}{\longrightarrow} \infty$ as $n \longrightarrow \infty$.
Now, remark that $K_{k,n}/K_{k',n} \stackrel{d}{=} W_{k,n}/W_{k',n}$. Rényi representation yields

$$K_{k,n}/K_{k',n} \stackrel{d}{=} T_k/T_{k'} = (k/k')[T_k/k]/[T_{k'}/k'] \stackrel{\text{a.s.}}{\longrightarrow} c.$$

The proof of $K_{k,n}/N_n \stackrel{\text{a.s.}}{\longrightarrow} 0$ as $n \longrightarrow \infty$ is similar.
Finally, Lemma 3 and Lemma 4 i) ii) imply that:

$$(1+Z_n)\frac{\varphi_\xi(1/k')}{\varphi_\xi(1/k)} \stackrel{\text{P}}{\longrightarrow} 1,$$

which concludes the proof using $k/k' \rightarrow c$.

iii) Since $1/K_{k,n} \stackrel{d}{=} W_{1,k-1} \stackrel{d}{=} 1-W_{k-1,k-1}$, and remarking that

$$k(1-W_{k-1,k-1}) \stackrel{d}{\longrightarrow} \text{Exp}(1),$$

it follows that $k/K_{k,n}$ has asymptotically a standard exponential distribution.
Now, let $E_1,\ldots,E_n$ be independent standard exponential random variables and let $E_{1,n} \leq \ldots \leq E_{n,n}$ be the corresponding order statistics. Dekkers and de Haan ([8], Lemma 2.1) shows that

$$\sqrt{\frac{k}{c-1}}(E_{n-k/c+1,n} - E_{n-k+1,n} - \ln(c)) \stackrel{d}{\longrightarrow} \mathcal{N}(0,1).$$

Since $K_{k,n} \stackrel{d}{=} \exp(E_{n,n} - E_{n-k+1,n})$ and $K_{k',n} \stackrel{d}{=} \exp(E_{n,n} - E_{n-k/c+1,n})$, the δ-method applied to the function $\varphi_\xi(\mathrm{e}^{-x})$ concludes the proof. ■

Proof of Lemma 5 – This proof is inspired by the one of [17], Lemma 0.13. We only give the proof for $\xi > 0$ (the case $\xi = 0$ is quite similar). For $j \in \mathbb{N}$, let

$$\beta_j = \exp\left[\frac{j}{\xi}\ln(1+\xi)\right].$$

Using relation (2), we have that for any $\varepsilon \in]0, 1-(1+\xi)^{-1}[$ there exist t_0 such that for all $t \geq t_0$ and for all $j \in \mathbb{N}\backslash\{0\}$,

$$1-\varepsilon \leq \frac{U(t\beta_j)-U(t\beta_{j-1})}{a(t\beta_{j-1})} \leq 1+\varepsilon. \tag{21}$$

Furthermore, since $a(.)$ is regularly varying at infinity with index ξ (see [17], Proposition 0.8 v) and Proposition 0.12), we have that for any $\varepsilon \in]0, 1-(1+\xi)^{-1}[$ there exist t_0 such that for all $t \geq t_0$ and for all $j \in \mathbb{N}\backslash\{0\}$,

$$\beta_1^{\xi}(1-\varepsilon) = (1+\xi)(1-\varepsilon) \leq \frac{a(t\beta_j)}{a(t\beta_{j-1})} \leq .(1+\xi)(1+\varepsilon). \tag{22}$$

Let $N \in \mathbb{N}$. We have:

$$\frac{U(t\beta_N)-U(t)}{a(t)} = \sum_{j=1}^{N} \frac{U(t\beta_j)-U(t\beta_{j-1})}{a(t\beta_{j-1})}\frac{a(t\beta_{j-1})}{a(t)}.$$

Using (21), we find that for $t \geq t_0$:

$$(1-\varepsilon)\sum_{j=1}^{N}\frac{a(t\beta_{j-1})}{a(t)} \leq \frac{U(t\beta_N)-U(t)}{a(t)} \leq (1+\varepsilon)\sum_{j=1}^{N}\frac{a(t\beta_{j-1})}{a(t)}. \tag{23}$$

Remarking that

$$\frac{a(t\beta_{j-1})}{a(t)} = \prod_{i=1}^{j-1}\frac{a(t\beta_i)}{a(t\beta_{i-1})},$$

(22) and (23) imply that there exist $\tilde{\beta_1}, \tilde{\beta_2} > 0$ such that for $t \geq t_0$:

$$\tilde{\beta_1}[(1+\xi)(1-\varepsilon)]^N - \tilde{\beta_1} \leq \frac{U(t\beta_N)-U(t)}{a(t)} \leq \tilde{\beta_2}[(1+\xi)(1+\varepsilon)]^N - \tilde{\beta_2}. \tag{24}$$

Let $N_x = [\xi/\ln(1+\xi)]\ln(x)$ (i.e. $x = \exp[N_x/\xi\ln(1+\xi)]$). Remarking that $\lfloor N_x \rfloor \leq N_x \leq \lfloor N_x \rfloor + 1$ implies that $\beta_{\lfloor N_x \rfloor} \leq x \leq \beta_{\lfloor N_x \rfloor+1}$. Since U is a non-decreasing function, we find that

$$\frac{U(t\beta_{\lfloor N_x \rfloor})-U(t)}{a(t)} \leq \frac{U(tx)-U(t)}{a(t)} \leq \frac{U(t\beta_{\lfloor N_x \rfloor+1})-U(t)}{a(t)},$$

and, using (24),

$$\tilde{\beta_1}[(1+\xi)(1-\varepsilon)]^{\lfloor N_x \rfloor} - \tilde{\beta_1} \leq \frac{U(tx)-U(t)}{a(t)} \leq \tilde{\beta_2}[(1+\xi)(1+\varepsilon)]^{\lfloor N_x \rfloor+1} - \tilde{\beta_2}.$$

Thus, there exist β_1 and β_2 such that:

$$\beta_1[(1+\xi)(1-\varepsilon)]^{N_x} - \tilde{\beta_1} \le \frac{U(tx)-U(t)}{a(t)} \le \beta_2[(1+\xi)(1+\varepsilon)]^{N_x} - \tilde{\beta_2}. \tag{25}$$

Remarking that

$$[(1+\xi)(1-\varepsilon)]^{N_x} = x^{\xi-\eta_1},$$

with $\eta_1 = -\xi\ln(1-\varepsilon)/\ln(1+\xi) \in]0,\xi[$ and

$$[(1+\xi)(1+\varepsilon)]^{N_x} = x^{\xi+\eta_2},$$

with $\eta_2 = \xi\ln(1+\varepsilon)/\ln(1+\xi) > 0$ concludes the proof. ■

Proof of Lemma 7 – The first step of the proof consists in establishing the following expansion:

$$-\frac{\varphi_\xi(1/c)}{Z_n} = \varphi_\xi(K_{k',n})\left[1-\frac{\sigma}{\sqrt{k}\varphi_\xi(K_{k',n}/K_{k,n})}Y_n\right]\left[1+o_P\left(\frac{1}{\varphi_\delta(k)}\right)\right]. \tag{26}$$

Let $V(t) = U(e^t)$. From Lemma 4 i), we have:

$$Z_n \stackrel{d}{=} \underbrace{\frac{V'[\ln(N_n/K_{k',n})]}{U(N_n/K_{k',n})-U(N_n)}}_{Z_{1,n}} \times \underbrace{\frac{U(N_n/K_{k,n})-U(N_n/K_{k',n})}{V'[\ln(N_n/K_{k',n})]}}_{Z_{2,n}}.$$

Clearly,

$$-\frac{1}{Z_{1,n}} = \int_0^{\ln(K_{k',n})} \frac{V'[\ln(N_n/K_{k',n})+s]}{V'[\ln(N_n/K_{k',n})]}ds,$$

and conditions $(H1)$ and $(H2)$ imply that uniformly on $[0,\ln(K_{k',n})]$,

$$\frac{V'[\ln(N_n/K_{k',n})+s]}{V'[\ln(N_n/K_{k',n})]} = e^{\xi s}\left[1+o_P\left(\frac{1}{\varphi_\delta(k)}\right)\right].$$

Thus,

$$-\frac{1}{Z_{1,n}} = \left[1+o_P\left(\frac{1}{\varphi_\delta(k)}\right)\right]\int_0^{\ln(K_{k',n})} e^{\xi s}ds = \varphi_\xi(K_{k',n})\left[1+o_P\left(\frac{1}{\varphi_\delta(k)}\right)\right]. \tag{27}$$

The proof of

$$Z_{2,n} = \varphi_\xi(K_{k',n}/K_{k,n})\left[1+o_P\left(\frac{1}{\varphi_\delta(k)}\right)\right], \tag{28}$$

follows the same lines. Collecting (27) and (28) yields

$$-\frac{\varphi_\xi(K_{k',n}/K_{k,n})}{Z_n} = \varphi_\xi(K_{k',n})\left[1+o_P\left(\frac{1}{\varphi_\delta(k)}\right)\right],$$

which proves (26) by remarking that

$$\frac{\varphi_\xi(1/c)}{\varphi_\xi(K_{k',n}/K_{k,n})} = 1-\frac{\sigma}{\sqrt{k}\varphi_\xi(K_{k',n}/K_{k,n})}Y_n.$$

Now, remark that (26) can be rewritten $v_n = u_n(1+\varepsilon_n)$ with $u_n = \varphi_\xi(K_{k',n})$, $v_n = -\varphi_\xi(1/c)/Z_n$ and

$$\varepsilon_n = \frac{-\sigma}{\sqrt{k}\varphi_\xi(K_{k',n}/K_{k,n})} Y_n + o_{\mathrm{P}}\left(\frac{1}{\varphi_\delta(k)}\right) = O_{\mathrm{P}}\left(k^{-1/2}\right) + o_{\mathrm{P}}\left(\frac{1}{\varphi_\delta(k)}\right) \xrightarrow{\mathrm{P}} 0,$$

from Lemma 4 iii). The second step of the proof is dedicated to the study of $\varphi_\xi^*\left(-\varphi_\xi(1/c)/Z_n\right)$. Five cases have to be considered:

If $\xi > 0$, since $u_n \xrightarrow{\mathrm{P}} +\infty$ as $n \to \infty$ (Lemma 4 iii)), Lemma 6 III) entails $\varphi_\xi^*(u_n) \overset{\mathrm{P}}{\sim} \varphi_\xi^*(v_n)$, i.e.,

$$\varphi_\xi^*\left(-\frac{\varphi_\xi(1/c)}{Z_n}\right) \overset{\mathrm{P}}{\sim} \varphi_\xi^*[\varphi_\xi(K_{k',n})] = (K_{k',n})^{\xi+\mathbb{I}\{\xi=0\}}, \tag{29}$$

by Lemma 6 I).

If $\xi = 0$, we have $u_n \xrightarrow{\mathrm{P}} +\infty$ as $n \to \infty$ and $u_n - v_n = u_n\varepsilon_n = o_{\mathrm{P}}[\ln(K_{k',n})\ln(k)] = o_{\mathrm{P}}(1)$. Thus, Lemma 6 II) implies (29).

If $-1/2 < \xi < 0$, we have from Lemma 4 ii) that $u_n \xrightarrow{\mathrm{P}} -1/\xi$ as $n \to \infty$. Remarking that $\varepsilon_n = o_{\mathrm{P}}[1/\varphi_\delta(k)] = o_{\mathrm{P}}(k^\xi)$ implies

$$\frac{\varepsilon_n}{\varphi_\xi^*(u_n)} = o_{\mathrm{P}}[(k/K_{k',n})^\xi] = o_{\mathrm{P}}(1),$$

which entails (29) by Lemma 6 IV) i).

If $\xi = -1/2$, we have $u_n \xrightarrow{\mathrm{P}} -1/\xi$ as $n \to \infty$. Remarking that $\varepsilon_n \overset{\mathrm{P}}{\sim} -\sigma/[\sqrt{k}\varphi_\xi(1/c)]Y_n$ yields:

$$\frac{\varepsilon_n}{\varphi_\xi^*(u_n)} \overset{\mathrm{P}}{\sim} -\frac{\sigma c^{-1/2}}{\varphi_\xi(1/c)} Y_n \sqrt{\frac{K_{k',n}}{k'}} = \alpha_n,$$

where α_n does not converges in probability to ∞ or 1 as $n \to \infty$ (see Lemma 4 iii)). Thus, from Lemma 6 IV) i), we have $\varphi_\xi^*(u_n) \overset{\mathrm{P}}{\sim} \varphi_\xi^*(v_n)(1-\alpha_n)$ i.e.

$$\varphi_\xi^*\left(-\frac{\varphi_\xi(1/c)}{Z_n}\right) \overset{\mathrm{P}}{\sim} (K_{k',n})^{-1/2}\left(1 + \frac{\sigma c^{-1/2}}{\varphi_\xi(1/c)} Y_n \sqrt{\frac{K_{k',n}}{k'}}\right) = k^{-1/2}T_n. \tag{30}$$

If $\xi < -1/2$, we have $u_n \xrightarrow{\mathrm{P}} -1/\xi$, $\varepsilon_n \overset{\mathrm{P}}{\sim} -\sigma/[\sqrt{k}\varphi_\xi(1/c)]Y_n$ and $\varepsilon_n/\varphi_\xi^*(u_n) \xrightarrow{\mathrm{P}} \infty$ when $n \to \infty$. Thus, Lemma 6 IV) ii) implies that $\varphi_\xi^*(v_n) \overset{\mathrm{P}}{\sim} -\varepsilon_n$ i.e.

$$\varphi_\xi^*\left(-\frac{\varphi_\xi(1/c)}{Z_n}\right) \overset{\mathrm{P}}{\sim} \frac{\sigma k^{-1/2}}{\varphi_\xi(1/c)} Y_n. \tag{31}$$

Lemma 4 iii) and (29)-(31) conclude the proof. ■

References

[1] B.C. Arnold, N. Balakrishnan, and H.N. Nagaraja. *A first course in order statistics*. Wiley and sons, 1992.

[2] J. Beirlant, P. Vynckier, and J.M. Teugels. Excess functions and estimation of the extreme value index. *Bernoulli*, **2**:293–318, 1996.

[3] N.H. Bingham, C.M. Goldie, and J.L. Teugels. *Regular Variation*. Cambridge University Press, 1987.

[4] S. Csörgo and D.M. Mason. Central limit theorems for sums of extreme values. *Mathematical Proceedings of the Cambridge Philosophical Society*, **98**:547–558, 1985.

[5] S Csörgo and L. Viharos. *Estimating the tail index*. Asymptotic methods in Probability and Statistics. B. Szyszkowicz, ed., North-Holland, Amsterdam, 1998. pp. 833-881.

[6] R. Davis and S.I. Resnick. Tail estimates motivated by extreme value theory. *Annals of Statistics*, **12**(4):1467–1487, 1984.

[7] L. de Haan. Slow variation and characterization of domains of attraction. In J. Tiago de Oliveira, editor, *Statistical Extremes and Applications*, pages 31–48. Reidel, Dorchrecht, 1984.

[8] A.L.M. Dekkers and L. de Haan. On the estimation of the extreme value index and large quantile estimation. *Annals of Statistics*, **17**:1795–1832, 1989.

[9] A.L.M. Dekkers, J.H.J. Einmahl, and L. de Haan. A moment estimator for the index of an extreme-value distribution. *Annals of Statistics*, **17**:1833–1855, 1989.

[10] P. Embrechts, C. Klüppelberg, and T. Mikosch. *Modelling Extremal Events for Insurance and Finance*. Springer-Verlag, 1997.

[11] A. Feueverger and P. Hall. Estimating a tail exponent by modelling departure from a Pareto distribution. *Annals of Statistics*, **27**:760–781, 1999.

[12] L. Gardes and S. Girard. Asymptotic distribution of a Pickands-type estimator of the extreme value index. *Comptes-Rendus de l'Acad¥mie des Sciences*, **t. 341**, Série I:53–58, 2005.

[13] B. Gnedenko. Sur la distribution limite du terme maximum d'une série aléatoire. *Annals of Mathematics*, **44**:423–453, 1943.

[14] P. Hall. On some simple estimates of an exponent of regular variation. *Journal of the Royal Statistical Society Series B*, **44**:37–42, 1982.

[15] B.M. Hill. A simple general approach to inference about the tail of a distribution. *Annals of Statistics*, **3**:1163–1174, 1975.

[16] J. Pickands III. Statistical inference using extreme-order statistics. *Annals of Statistics*, **3**:119–131, 1975.

[17] S.I. Resnick. *Extreme Values, regular Variation, and Point Process*. Springer-Verlag, New-York, 1987.

In: Focus on Probability Theory
Editor: Louis R. Velle, pp. 127-134

ISBN 1-59454-474-3

Chapter 7

On some Moments of the Limit Random Variable for a Normalized Supercritical Galton-Watson Process

Aleksander M. Iksanov
Faculty of Cybernetics, Kiev National University
01033 Kiev, Ukraine

Abstract

Let $Z_n, n = 1, 2, \dots$ be the supercritical Galton-Watson process with the finite mean family size. Let f be a nonnegative, nondecreasing and concave on $[0, \infty)$ function. For $B, d \geq 0$ a new function g is defined by $g(x) := B + \int_d^x (f(y)/y)dy$ for $x \geq d$, and $g(x) = 0$ for $x < d$. Assume that, when $n \to \infty$, $Z_n/\mathbb{E}Z_n$ converges almost surely and in mean to a random variable W. By using a new (nonanalytic) approach it is proved that $\mathbb{E}Wf(W) < \infty$ if and only if $\mathbb{E}Z_1 g(Z_1) < \infty$.

Key words: Galton-Watson process; moments; random power series; change of measure.

1 Introduction and Results

Let $Z_n, n = 1, 2, \dots$ be the supercritical Galton-Watson process with the mean family size $m := \mathbb{E}Z_1 \in (1, \infty)$. By the Kesten-Stigum theorem ([4], theorem I.10.1) the nonnegative martingale $W_n := Z_n/\mathbb{E}Z_n$ converges almost surely and in mean to a random variable W with $\mathbb{E}W = 1$ if and only if

$$\mathbb{E}Z_1 \log(1 + Z_1) < \infty. \tag{1}$$

Moreover, almost surely $\{\omega : W(\omega) > 0\} = \{\omega : Z_n(\omega) \to \infty\}$.

Let $f, g \geq 0$ be some measurable, unbounded near infinity functions. Assume these are functionally dependent in a sense. In the literature there are some results [3, 6, 1] which investigate the relation between the existence of f-moments of W and g-moments of Z_1. In particular, the equivalence

$$\mathbb{E}W^p < \infty \Leftrightarrow \mathbb{E}Z_1^p < \infty, p > 1 \tag{2}$$

is well-known and can be proved in a number of ways. The first, more interesting result was obtained in [3]:

$$\mathbb{E}W\log^p(1+W)<\infty \Leftrightarrow \mathbb{E}Z_1\log^{p+1}(1+Z_1)<\infty, p>0. \tag{3}$$

Let C^* be a class of convex and differentiable functions ϕ which have the following properties (1) $\phi(0)=\phi'(0)=0$; (2) for some $n\in\mathbb{N}$ $\phi(x^{2^{-n}})$ is a convex function with concave derivative; (3)$\mathbb{L}\phi(x):=\int_0^x\int_0^s(\phi'(r)/r)dr<\infty$. The class C^* was considered in [1]. In the same paper the following equivalence

$$0<\mathbb{E}\phi(W)<\infty \Leftrightarrow \mathbb{E}\mathbb{L}\phi(Z_1)<\infty, \phi\in C^*$$

was proved.

Let us introduce **condition A**: function $f\geq 0$ is nondecreasing and concave on $[0,\infty)$; $\lim_{x\to\infty} f(x)=\infty$. For fixed $B,d\geq 0$ a new function g is defined by

$$g(x):=B+\int_d^x (f(y)/y)dy, \text{ for } x\geq d; g(x)=0 \text{ for } x<d.$$

The main result of this paper is given next.

Proposition 1.1. If the conditions **A** and (1) hold, then

$$\mathbb{E}Wf(W)<\infty \Leftrightarrow \mathbb{E}Z_1 g(Z_1)<\infty. \tag{4}$$

Remark. Define

$$\mathcal{D}^*:=\{t: t(x):=xf(x), f(x) \text{ satisfies the condition } \mathbf{A}\}.$$

Then $\mathcal{D}^*\cap C^*\neq\emptyset$, but neither $\mathcal{D}^*$ comprises C^*, nor C^* comprises $\mathcal{D}^*$.

The equivalence (2) with $p\in(1,2)$ suggests that sometimes a function g (in Proposition 1.1) can be replaced with an f. On the other hand, the equivalence (3) reveals that this is not always possible. The following question arises: does there exist concave functions f which are not power functions for which

$$\mathbb{E}Wf(W)<\infty \Leftrightarrow \mathbb{E}Z_1 f(Z_1)<\infty? \tag{5}$$

Corollary 1.2. Assume that f satisfies the condition **A**, and for $x\geq d$ it takes the form $f(x)=x^\alpha h(x)$, where $\alpha\in(0,1)$ and $h(x)$ does not decrease for $x\geq d$. Then (5) holds.

The results of [3, 6] were obtained by using analytic methods. In [1] a probabilistic approach was proposed. It is based on using several convex function inequalities for martingales. The technique developed in our paper generalizes an approach proposed in the proof of Proposition 4 [8] (a similar approach has independently been used in [7] in another context). Note that in [8] a more general process (branching random walk) was considered, but f-moments were only studied for the power functions $f(x)=x^p, p\in(1,2)$.

The method of this work relies upon an appropriate change of measure proposed in [9] (move from a measure GW to a measure $\widehat{GW}$ is described in Section 2). Under $\widehat{GW}$, the random variable W can be "compared" to a random (convergent) power series, the study of which is rather simple. Our proof uses Jensen's inequality for conditional expectations and subadditivity of f. Thus the concavity of f is an essential (indispensable?) requirement for our method to work.

All proofs are contained in Section 2. In Section 3 some explicit examples of f and g are given that illustrate Proposition 1.1 and Corollary 1.2.

2 Proofs

As $f(x)$ is nonnegative and concave, $f(x)/x$ does not increase. Also, for $x > 0$ the right derivative f' is defined and does not increase. For later use, let us note that the concave function f is absolutely continuous, and f' is a version of its density. Thus, if f' is under the integral sign, it can be treated as the usual (two-sided) derivative.

Proof of Corollary 1.2. According to what has been said in the previous paragraph, we have for $x > d$

$$f'(x) \leq f(x)/x = g'(x), \quad f(x) \leq g(x) + f(d) - g(d). \tag{6}$$

Therefore, $\mathbb{E}Z_1 g(Z_1) < \infty$ always implies $\mathbb{E}Z_1 f(Z_1) < \infty$ for any functions f which satisfy the condition **A** (at this point we do not require that the function $x^{-\alpha}f(x)$ be nondecreasing). On the other hand, an appeal to the explicit form of f gives

$$g(x) = g(d) + \int_d^x \frac{f(u)}{u} du \leq \alpha^{-1} f(x) - \alpha^{-1} d^{\alpha} h(x) + g(d).$$

Hence, $\mathbb{E}Z_1 f(Z_1) < \infty$ implies $\mathbb{E}Z_1 g(Z_1) < \infty$ for $f(x) = x^{\alpha}h(x)$. A combination of the two implications allows us to conclude that the finiteness of these moments are in fact equivalent. It remains to use Proposition 1.1.

Lemma 2.1 given next is related to Theorem 2 [10]. However, our assumptions and proof are distinct.

Lemma 2.1. Let $b \in (0,1)$ be fixed, and $\xi_1, \xi_2, \dots$ be independent copies of a nonnegative random variable ξ such that

$$\mathbb{E}\log(\xi + 1) < \infty. \tag{7}$$

Set $X := \sum_{n=0}^{\infty} b^n \xi_n$. Then $X < \infty$ almost surely. If functions f and g verify the condition **A**, then $\mathbb{E}f(X) < \infty$ is implied by $\mathbb{E}g(\xi) < \infty$.

Proof. Almost sure finiteness of X is well-known and follows, for example, by the Cauchy's root test for convergence of series and the Borel-Cantelli lemma. The concavity of $f \geq 0$ implies its subadditivity, i.e. $f(x+y) \leq f(x) + f(y), x, y \geq 0$. For any $k \in \mathbb{N}$ and fixed $c > f(d)$

$$\mathbb{E}\left(f\left(\sum_{n=0}^{k} b^n \xi_n \right) \vee c \right) \leq \mathbb{E} \sum_{n=0}^{k} (f(b^n \xi_n) \vee c) = \sum_{n=0}^{k} \int_c^{\infty} \mathbb{P}\{f(b^n \xi_n) > x\} dx =: I.$$

Change of variable $v = b^{-n} f^{-1}(x)$ and using the inequality

$$f(v) \geq \int_{b^{k+1}v}^{v} f'(y) dy = \sum_{n=0}^{k} \int_{b^{n+1}v}^{b^n v} f'(y) dy \geq (1-b) v \sum_{n=0}^{k} b^n f'(b^n v),$$

gives

$$I = \sum_{n=0}^{k} \int_{b^{-n}d}^{\infty} \mathbb{P}\{\xi_n > v\} b^n f'(b^n v) dv \leq \int_d^{\infty} \mathbb{P}\{\xi > v\} \left(\sum_{n=0}^{k} b^n f'(b^n v) \right) dv$$

$$\leq (1-b)^{-1} \int_d^{\infty} \mathbb{P}\{\xi > v\} (f(v)/v) dv = (1-b)^{-1} \int_d^{\infty} \mathbb{P}\{\xi > v\} g'(v) dv.$$

Thus, $\mathbb{E}\left(f\left(\sum_{n=0}^{k} b^n \xi_n\right) \vee c\right)$ is bounded from the above by a constant which does not depend on k. It remains to apply the Lebesgue monotone convergence theorem and pass to the limit as $k \to \infty$.

Proof of Proposition 1.1. Let $Y_1, Y_2, \ldots$ be independent copies of a random variable Y with the distribution $\mathbb{P}\{Y = k\} = kp_k/m, k = 1,2,\ldots,$ where $p_k, k = 0,1,\ldots$ is the family size distribution. Let GW be a probability measure on the set of the Galton-Watson trees. In [9] a probability measure $\widehat{GW}$ on the set of *size-biased* trees was constructed. Each such a tree is defined as follows. An initial ancestor v_0 (root of the tree) gives birth to Y_1 children. While one of these children v_1 chosen at random gives birth to Y_2 children, her sisters are the roots of independent Galton-Watson trees which are also independent of Y_2. Pick one of the children of v_1, call it v_2. She gives birth to Y_3 children and so on.

Sufficiency. Assume that $\mathbb{E}_{GW} Z_1 g(Z_1) < \infty$. This inequality is equivalent to the following

$$\mathbb{E}_{\widehat{GW}} g(Y-1) < \infty. \tag{8}$$

Let $\mathcal{G}$- be the σ-field generated by $Y_1, Y_2, \ldots$ From the results of [9] it follows that

$$\mathbb{E}_{\widehat{GW}}(W_n | \mathcal{G}) = \sum_{k=1}^{n} (Y_k - 1)/m^k, \tag{9}$$

and that W_n converges almost surely to W. Therefore, by Fatou's lemma

$$\mathbb{E}_{\widehat{GW}}(W | \mathcal{G}) \le \sum_{k=1}^{\infty} (Y_k - 1)/m^k =: V. \tag{10}$$

Since f is nondecreasing and concave, then using Jensen's inequality and applying the expectation operator one more results in

$$\mathbb{E}_{\widehat{GW}}(f(W) | \mathcal{G}) \le f(\mathbb{E}_{\widehat{GW}}(W | \mathcal{G})) \le f(V);$$

$$\mathbb{E}_{\widehat{GW}} f(W) \le \mathbb{E}_{\widehat{GW}} f(V). \tag{11}$$

According to Lemma 2.1, the condition (8) ensures the finiteness of the right-hand side of (11). Hence, we have

$$\mathbb{E}_{GW} W f(W) = \mathbb{E}_{\widehat{GW}} f(W) < \infty.$$

This completes the proof of sufficiency of Proposition 1.1.

To prove Proposition 1.1 in the reverse direction we need three lemmas. Lemma 2.2 proposes a $\widehat{GW}$-counterpart of the inequality obtained in Lemma II.2.6. [2] which deals with distributions of W and $\sup_n W_n$ under GW. One step of our proof of Lemma 2.2 uses an idea of the proof of Lemma 2 [5], where a branching random walk generalization of the inequality from [2] was found.

Lemma 2.2. For each $a > 0$ small enough there exists $B > 1$ such that whenever $t > 1$ the following inequalities hold

$$\widehat{GW}\{W > t\} \le \widehat{GW}\{\sup_n W_n > t\} \le B\widehat{GW}\{W > at\}.$$

In the sequel, 1_A denotes the indicator of the event A. To prove Lemma 2.2 we need an auxiliary result.

Lemma 2.3. Let $\xi_1, \xi_2, \ldots$ be independent copies of a nonnegative random variable ξ. Take any $b > 0$ such that $\mathbb{P}\{\xi > b\} > 0$. For any $k \in \mathbb{N}$ and any $a < b\mathbb{P}\{\xi > b\}$,

$$\mathbb{P}\{\frac{1}{k}\sum_{i=1}^{k}\xi_i > a\} \geq \frac{(\mathbb{E}(\xi \wedge b) - a)^2}{\mathbb{E}(\xi \wedge b - a)^2}. \tag{12}$$

Proof. We will make use of a standard inequality: if η is a random variable with $\mathbb{E}\eta \geq 0$, then

$$\mathbb{P}\{\eta > 0\} \geq \frac{(\mathbb{E}\eta)^2}{\mathbb{E}\eta^2}. \tag{13}$$

Set $\zeta_i := \xi_i \wedge b - a$ and $\eta_k := \frac{1}{k}\sum_{i=1}^{k}\zeta_i$. Since $\mathbb{E}\eta_k = \mathbb{E}\zeta_1 > 0$, (13) applies.

$$\mathbb{P}\{\frac{1}{k}\sum_{i=1}^{k}\xi_i > a\} \geq \mathbb{P}\{\eta_k > 0\} \geq \frac{(\mathbb{E}\eta_k)^2}{\mathbb{E}\eta_k^2} = \frac{k(\mathbb{E}\zeta_1)^2}{\mathbb{E}\zeta_1^2 + (k-1)(\mathbb{E}\zeta_1)^2} \geq \frac{(\mathbb{E}\zeta_1)^2}{\mathbb{E}\zeta_1^2},$$

the latter inequality being implied by the fact that the function $x \longrightarrow \frac{x(\mathbb{E}\zeta_1)^2}{\mathbb{E}\zeta_1^2 + (x-1)(\mathbb{E}\zeta_1)^2}$ does not decrease for $x \geq 1$. The proof of Lemma 2.3 is complete.

Proof of Lemma 2.2. The left-hand side inequality is obvious. Let us prove the rest. Under GW, the following well-known equality of distribution holds

$$W \stackrel{d}{=} \frac{1}{m^n}\sum_{i=1}^{Z_n} W^{(i)},$$

where given Z_n $W^{(1)}, W^{(2)}, \ldots$ are (conditionally) independent copies of W. Under $\widehat{GW}$,

$$W \stackrel{d}{=} \frac{1}{m^n}W + \frac{1}{m^n}\sum_{i=1}^{Y_n} V^{(i)}, \tag{14}$$

where given Y_n $V^{(1)}, V^{(2)}, \ldots$ are (conditionally) independent copies of a random variable V with distribution $\widehat{GW}\{V \leq x\} = GW\{W \leq x\}$. Consider the events $E_n := \{\max_{0 \leq i \leq n-1} W_i \leq t, W_n > t\}$. Pick any $b > 0$ and take any $a \in (0, bGW\{W > b\})$. For such a we have almost surely

$$\widehat{GW}\{\frac{1}{Y_n}\sum_{i=1}^{Y_n} V^{(i)} > a|\sigma(Y_0, \ldots, Y_n)\} \stackrel{(12)}{\geq} \frac{(\mathbb{E}_{\widehat{GW}}(V \wedge b) - a)^2}{\mathbb{E}_{\widehat{GW}}(V \wedge b - a)^2} =: (1/B).$$

Therefore

$$\widehat{GW}\{W > at|E_n\} \stackrel{(14)}{\geq} \widehat{GW}\{\frac{1}{W_n m^n}\sum_{i=1}^{Y_n} V^{(i)} > \frac{at}{W_n}|E_n\} \geq$$

(use $W_n \leq Y_n m^{-n}$ under $\widehat{GW}$)

$$\geq \widehat{GW}\{\frac{1}{Y_n}\sum_{i=1}^{Y_n} V^{(i)} > a|E_n\} \geq 1/B.$$

The inequality

$$\widehat{GW}\{W > at\} \geq \sum_{n=1}^{\infty} \widehat{GW}\{W > at|E_n\}\widehat{GW}\{E_n\} \geq (1/B)\widehat{GW}\{\sup_n W_n > t\}$$

completes the proof of Lemma 2.2.

The next lemma is an elementary consequence of Lemma 2.2.

Lemma 2.4. For any nonnegative, nondecreasing and anti-starshaped (in particular, concave) function h,

$$\mathbb{E}_{\widehat{GW}} h(W) < \infty \Leftrightarrow \mathbb{E}_{\widehat{GW}} h(\sup_n W_n) < \infty.$$

Proof of Proposition 1.1. *Necessity.* Assume that $\mathbb{E}_{\widehat{GW}} f(W) = \mathbb{E}_{GW} W f(W) < \infty$. Since, under $\widehat{GW}$, $Z_n \geq Y_n$, and f does not decrease, then $f(\sup_n W_n) \geq f(\sup_n Y_n/m^n)$. Thus,

$$\infty > \mathbb{E}_{\widehat{GW}} f(\sup_n W_n) \geq \mathbb{E}_{\widehat{GW}} f(\sup_n Y_n/m^n),$$

the first inequality being implied by Lemma 2.4.

If the support of a distribution of Y is bounded, then for any $r \in \mathbb{N}$ $\mathbb{E}_{\widehat{GW}} Y^r = m^{-1}\mathbb{E}_{GW} Z_1^{r+1} < \infty$. Consequently, (2) implies $\mathbb{E}_{GW} W^{r+1} < \infty$, and the result of the proposition is trivial. Hence in what follows we assume that the support of a distribution of Y is unbounded from the above. In that case there exists an $s > d$ such that $\widehat{GW}\{Y > t\} > 0$ for all $t \geq s$.

In view of (1), $\mathbb{E}_{\widehat{GW}} \log(Y+1) < \infty$. Therefore, the infinite product $\prod_{j=1}^{\infty} \widehat{GW}\{Y \leq m^j s\} =: C$ is non-zero. Furthermore, we have for $t \geq s$

$$\widehat{GW}\{\sup_n (Y_n/m^n) > t\} = \sum_{n=1}^{\infty} \widehat{GW}\{Y > m^n t\} \prod_{j=1}^{n-1} \widehat{GW}\{Y \leq m^j t\} \geq \tag{15}$$

$$\geq \sum_{n=1}^{\infty} \widehat{GW}\{Y > m^n t\} \prod_{j=1}^{n-1} \widehat{GW}\{Y \leq m^j s\} \geq C \sum_{n=1}^{\infty} \widehat{GW}\{Y > m^n t\}.$$

To complete the proof it suffices to show that

$$\int_s^{\infty} g'(u)\widehat{GW}\{Y > mu\}du < \infty. \tag{16}$$

Recall that $g'(u) = u^{-1} f(u)$ for $u \geq s$. Now (16) follows from the inequalities

$$\infty > \int_s^{\infty} f'(t)\widehat{GW}\{\sup_n (Y_n/m^n) > t\}dt \overset{(15)}{\geq} C\int_s^{\infty} f'(t) \sum_{n=1}^{\infty} \widehat{GW}\{Y > m^n t\}dt \geq$$

$$\geq (m-1)^{-1} C \int_s^{\infty} f'(t) \int_{mt}^{\infty} z^{-1}\widehat{GW}\{Y > z\}dzdt =$$

$$= const + (m-1)^{-1} C \int_s^{\infty} t^{-1} f(t)\widehat{GW}\{Y > mt\}dt.$$

The proof of Proposition 1.1 is finished.

3 Examples

In this Section we give examples of some pairs f and g which satisfy the condition **A**. Also we point out corresponding equivalences like (4). Notice that we omit multiplicative constants and summands of lower order.
(1) $f(x) = x^{p-1}, p \in (1,2)$, $g(x) = (p-1)^{-1}x^{p-1}$ lead to (2) with $p \in (1,2)$;
(2) if $p \in (0,1]$ $f(x) = e^{-1}x1_{\{x\in[0,e)\}} + \log^p x 1_{\{x\geq e\}}$,
$g(x) = (p+1)^{-1}(\log^{p+1} x - 1)1_{\{x\geq e\}}$,
and if $p > 1$ $f(x) = p(p-1)e^{1-p}x1_{\{x\in[0,e^{p-1})\}} + (\log^p x + (p-1)^2)1_{\{x\geq e^{p-1}\}}$,
$g(x) = ((p+1)^{-1}(\log^{p+1} x - (p-1)^p) + (p-1)^2(\log x - p + 1))1_{\{x\geq e^{p-1}\}}$ lead to (3);
(3) $f(x) = \beta^\beta e^{-\beta-\exp(\beta)}x1_{\{x\in[0,\exp(e^\beta))\}} + (\log^\beta \log x + c)1_{\{x\geq \exp(e^\beta)\}}, \beta > 0$, where $c := (\beta/e)^\beta - \beta$,
$g(x) = (\log x \log^\beta \log x - \beta e^\beta - \beta \int_{e^\beta}^{\log x} \log^{\beta-1} u du + c(\log x - e^\beta))1_{\{x\geq e^\beta\}}$. Since if $\beta \in (0,1]$

$$(\log x - e^\beta)\log^{\beta-1}\log x \leq \int_{e^\beta}^{\log x} \log^{\beta-1} u du \leq \log x - e^\beta,$$

and if $\beta > 1$ the similar inequality holds with the left- and right-hand sides interchanged, we have

$$\mathbb{E}W \log^\beta \log(W+e) < \infty \Leftrightarrow \mathbb{E}Z_1 \log(Z_1+e)\log^\beta \log(Z_1+e) < \infty.$$

We now give an example that illustrates Corollary 1.2.
(4) $f(x) = bx1_{\{x\in[0,a)\}} + (x^\beta \log^\gamma x + c)1_{\{x\geq a\}}, \beta \in (0,1), \gamma > 0$, where $a := \exp(\gamma/(1-\beta))$, $b := a^{\beta-1}\log^{\gamma-1} a(\beta \log a + \gamma)$, $c := a^\beta \log^{\gamma-1} a(\gamma - (1-\beta)\log a)$, verifies the conditions of Corollary 1.2. Therefore,

$$\mathbb{E}W^{\beta+1}\log^\gamma(W+1) < \infty \Leftrightarrow \mathbb{E}Z_1^{\beta+1}\log^\gamma(Z_1+1) < \infty.$$

References

[1] Alsmeyer, G., Rösler, U., 2004. On the existence of ϕ-moments of the limit of a normalized supercritical Galton-Watson process. *J. Theor. Prob.* **17**, 905-928.

[2] Asmussen, S., Hering, H., 1983. *Branching processes*, Birkhäuser, Boston.

[3] Athreya, K.B., 1971. A note on a functional equation arising in Galton-Watson branching processes, *J.Appl.Prob.* **8**, 589-598.

[4] Athreya, K.B. and Ney, P., 1972. *Branching processes*, Springer, New York.

[5] Biggins, J.D., 1979. Growth rates in the branching random walk, *Z.Wahrsch. Gebiete.* **48**, 17-34.

[6] Bingham, N.H., Doney, R.A., 1974. Asymptotic properties of supercritical branching processes I: The Galton-Watson process, *Adv.Appl.Prob.* **6**, 711-731.

[7] Hardy, R., Harris, S.C. 2004. *Spine proofs for L_p-convergence of branching-diffusion martingales*. Mathematics Preprint. 0405, University of Bath. Available online at http://www.bath.ac.uk/~massch/Research/Papers/spine-Lp-cgce.pdf

[8] Iksanov, A.M., 2004. Elementary fixed points of the BRW smoothing transforms with infinite number of summands. *Stoch.Proc.Appl.* **114**, 27-50.

[9] Lyons, R., Pemantle, R., Peres, Y., 1995. Conceptual proofs of *LlogL* criteria for mean behaviour of branching processes, *Ann.Prob.* **23**, 1125-1138.

[10] Zerner, M.P.F., 2002. Integrability of infinite sums of heavy-tailed i.i.d. random variables. *Stoch.Proc.Appl.* **99**, 81-94.

In: Focus on Probability Theory
Editor: Louis R. Velle, pp. 135-169

ISBN 1-59454-474-3

Chapter 8

Partially Reflected Brownian Motion: A Stochastic Approach to Transport Phenomena

Denis S. Grebenkov*
Laboratoire de Physique de la Matiere Condensée,
Ecole Polytechnique 91128 Palaiseau Cedex, France

Abstract

The transport phenomena are known to be important for various scientific domains: examples can be found in physics, electrochemistry, heterogeneous catalysis, physiology, etc. To obtain new information about diffusive or Laplacian transport towards a semi-permeable or resistive interface, one can study the random trajectories of diffusive particles modelled, in a first approximation, by the partially reflected Brownian motion. This stochastic process turns out to be a convenient mathematical foundation for discrete, semi-continuous and continuous theoretical descriptions of the diffusive transport.

This paper presents an overview on these topics with a special emphasis on the close relations between stochastic processes with partial reflections and the Laplacian transport phenomena. We give selected examples of these phenomena followed by a brief introduction to the partially reflected Brownian motion and related probabilistic objects (e.g., local time process and spread harmonic measure). A particular attention is paid to a useful relation to the Dirichlet-to-Neumann operator. Some practical consequences and further perspectives are discussed.

Keywords: Diffusion with Reflections; Mixed Boundary Value Problems; Laplacian Transport Phenomena.

Introduction

An erratic motion of pollens of Clarkia (primrose family) discovered by Robert Brown in 1827 and quantitatively described by Albert Einstein in 1905 gave a substantial impact

*E-mail address: Denis.Grebenkov@polytechnique.fr

for development of mathematical theory of stochastic processes, an important branch of modern mathematics. Supported by rigorous mathematical foundations, the Brownian motion and related stochastic processes found numerous applications in different scientific domains, from theoretical physics to biology and economics. To study the transport of species diffusing from a remote source towards and across semi-permeable or resistive interface (e.g., cellular membrane), one can employ either an averaged description in terms of an appropriate boundary value problem for the concentration of species, or stochastic analysis of its random trajectories. In the first case, a finite permeability of the interface leads to the mixed boundary condition, while in the second case it can be modelled by partial reflections on the boundary. Physical or chemical processes governed by the Laplace equation (stationary diffusion) with mixed boundary condition are generally called *Laplacian transport phenomena*. Its examples are found in physiology (oxygen diffusion towards and across alveolar membranes), in electrochemistry (electric transport through metallic electrodes), in heterogeneous catalysis (diffusion of reactive molecules towards catalytic surfaces). The theoretical or numerical analysis of these phenomena is generally complicated by an irregular geometry of the interface (e.g., microroughness of metallic electrodes, see Section 1). Studying random trajectories of diffusing species, one can extract a more subtle information about the system in question.

In this paper, we present a particular stochastic process, called *partially re ected Brownian motion* (PRBM), and its application to study the Laplacian transport phenomena. The main purpose we are standing for is to capture the attention to this motion for understanding the influence of a geometrical irregularity of the interface on transport properties of the whole system. Bearing in mind the particular role of the geometry, we would like to "bridge" theoretical, numerical and experimental studies of the Laplacian transport phenomena on the one side, and powerful mathematical methods of stochastic analysis on the other side. Since the extensive literature existing on both topics is generally difficult to get through for non-specialists, we prefer to use a descriptive style of writing in order to give the whole vision of the problem, without specifying particular details which can be easily found anywhere else (e.g., see references at the end of this paper).

In the first section, we present three examples of the Laplacian transport phenomena found in different scientific fields. Their mathematical description by the mixed boundary value problem opens encouraging possibilities to apply powerful tools of potential theory, variational analysis and probability theory. The second section is devoted to remind some basic definitions of stochastic process theory: stopping times, reflected Brownian motion, local time process, harmonic measure, etc. In the third section, we introduce the partially reflected Brownian motion and show its properties for a planar surface. An important relation to the Dirichlet-to-Neumann operator is revealed and then illustrated by several examples. The last section contains the stochastic descriptions of the Laplacian transport phenomena: recently developed continuous approach and two other approaches are described. In the conclusion, we summarize the essential aspects of the paper.

1 Laplacian Transport Phenomena

The transport of species between two distinct "regions" separated by an interface occurs in various biological systems: water and minerals are pumped by roots from the earth, ions and biological species penetrate through cellular membranes, oxygen molecules diffuse towards alveolar ducts, and so on. The transport processes are relevant for many other scientific domains, for example, in heterogeneous catalysis (principal industrial process in petrochemistry) or in electrochemistry. In this section, we will give three important examples of particular transport process, called *Laplacian* or *diffusive transport*.

1.1 Stationary Diffusion across Semi-permeable Membranes

Let us begin by considering the respiration process of mammals. Inbreathing a fresh air, one makes it flow from the mouth to the dichotomic bronchial tree of the lungs (Fig. 1). For humans, first fifteen generations of this tree serve for convectional transport of the air towards pulmonary acini, terminal gas exchange units [1]. A gradual increase of the total cross-section of bronchiae leads to a decrease of air velocity. At the entrance of the acinus, it becomes lower than the characteristic diffusion velocity [2, 3]. As a consequence, one can describe the gas exchange into the acinus as stationary diffusion of oxygen molecules in air from the entrance ("source" with constant concentration C_0 during one cycle of respiration) to the alveolar membranes [4]. In the bulk, the flux density is proportional to the gradient of concentration (Fick's law), $\mathbf{J} = -D\nabla C$, where D is the diffusion coefficient. The mass conservation law written locally as div $\mathbf{J} = 0$ leads to the Laplace equation $\Delta C = 0$ in the bulk. The flux density *towards* the interface is simply $J_n = D\, \partial C/\partial n$, where the normal derivative $\partial/\partial n$ is directed to the bulk. Arrived to the alveolar membrane, oxygen molecules can penetrate across the boundary for further absorption in blood, or to be "bounced" on it and to continue the motion. The "proportion" of absorbed and reflected molecules can be characterized by *permeability* W varying from 0 (perfectly reflecting boundary) to infinity (perfectly absorbing boundary). Thus, the flux density *across* the alveolar membrane is proportional to the concentration, $J_n = WC$. Equating these two densities, one obtains a mixed boundary condition, $D(\partial C/\partial n) = WC$, called also *Fourier* or *Robin* boundary condition. Resuming these relations, one provides the following mathematical description for the diffusion stage of human or, in general, mammalian respiration:

$$\Delta C = 0 \qquad \text{in the bulk} \tag{1}$$

$$C = C_0 \qquad \text{on the source} \tag{2}$$

$$\left[I - \Lambda \frac{\partial}{\partial n}\right] C = 0 \qquad \text{on the alveolar membrane} \tag{3}$$

where the involved physics and physiology are characterized by a single parameter, $\Lambda = D/W$, being homogeneous to the length (I stands for the identity operator). Note also that the dependence on constant C_0 is irrelevant. In what follows, we address to this "classical" boundary value problem. The essential complication resides in a very irregular geometry of the pulmonary acinus which presents a branched structure of eight generations (for humans), "sticked" by alveolar ducts (Fig. 1). For small Λ, only a minor part of the boundary is involved to the transport process (so-called *Dirichlet active zone*), whereas the flux across the rest of the boundary is almost zero (this effect is called *diffusional screening* [4–10]). With an increase of Λ, larger and larger part of the boundary becomes active. As a result,

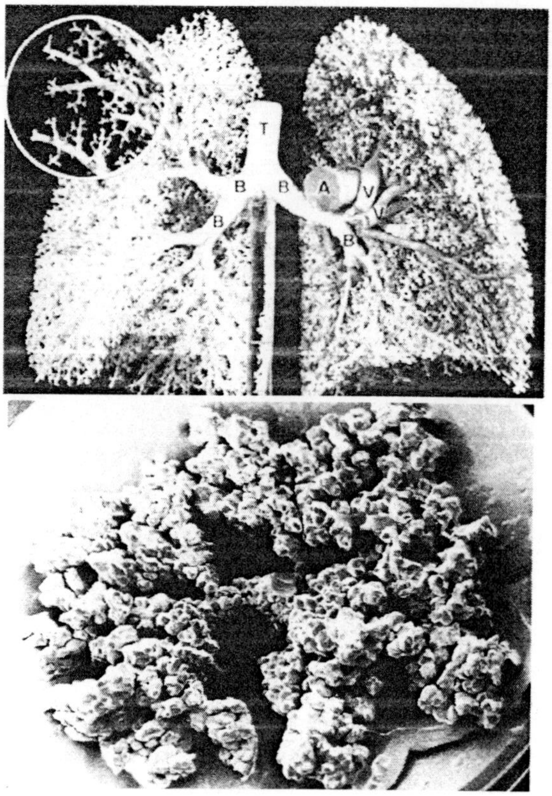

Figure 1: On the left, a cast of human lungs; on the right, a cast of pulmonary acinus [1, 11] (by E. Weibel).

the efficiency of human lungs depends on the parameter Λ in a non trivial manner that implies different physiological consequences. Considering the trajectory of a chosen oxygen molecule, one finds the Brownian motion from the source towards the alveolar membrane, with multiple bounces on the boundary and final absorption. One will see that such trajectories correspond to the partially reflected Brownian motion (Section 3). The profound study of the interplay between the irregular geometry of the acinus and erratic random motion of oxygen molecules inside should lead to a better understanding of the physiological functioning of human lungs.

1.2 Heterogeneous Catalysis

The similar description can be brought to the heterogeneous catalysis omnipresent at petrochemistry. One considers reactive molecules A injected into a solvent and then diffusing towards a catalytic surface. Hitting this boundary, they can be transformed into other molecules A^* with a finite reaction rate K, or to be bounced for further diffusion. The new molecules diffuse from the catalytic surface and then they are collected with the help of appropriate physical or chemical technics. Assuming the presence of a remote source of

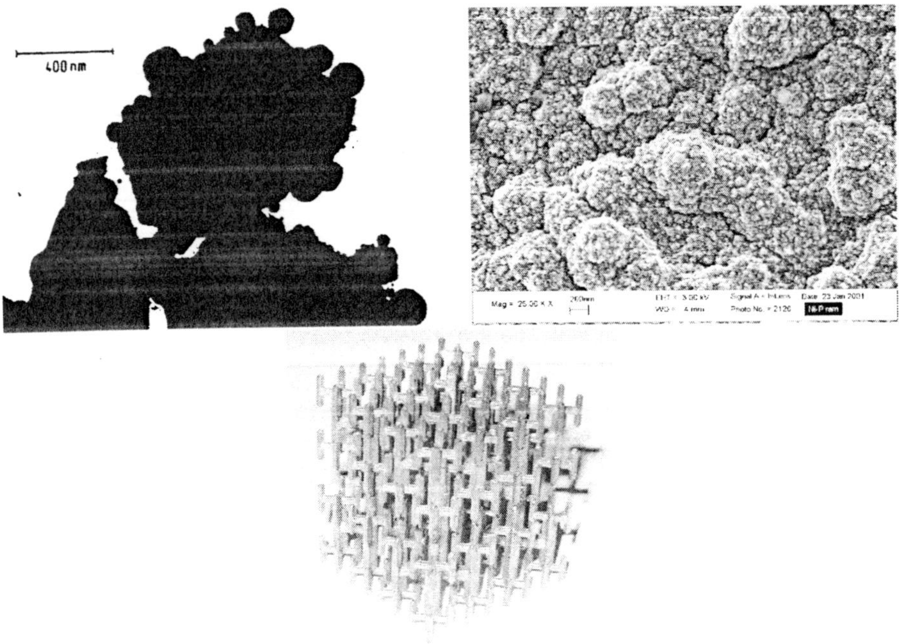

Figure 2: On the left, an example of an irregular catalytic surface (by J.S. Andrade jr.); at the center, photo of a rough metallic surface of nickel electrode (by E. Chassaing); on the right, photo of an irregular metallic electrode used to study the Laplacian transport phenomena experimentally (by B. Sapoval).

reactive molecules A, one can model, in a first approximation[1], the heterogeneous catalysis by the mixed boundary value problem (1–3) with a characteristic length $\Lambda = D/K$ [12–16]. The keynote of this similitude is related to the fact that each reactive molecule arrived to the boundary terminates its motion after a number of successive reflections. The mechanism leading to their termination is different: for the oxygen diffusion, the molecules are absorbed by the alveolar membrane, whereas for the heterogeneous catalysis, the reactive molecules are transformed into other molecules which do not participate to further process. Since the overall production of new molecules A^* depends on the total surface area of the catalytic surface, one tries to design catalysts with the largest possible surface (for given volume), realizing porous and very irregular boundaries (Fig. 2). As a consequence, the diffusional screening becomes important for understanding numerous industrial processes in petrochemistry. Since random trajectories of reactive molecules correspond to the partially reflected Brownian motion, its study may allow to design more efficient catalysts.

[1]This description is probably too simplified in order to model the heterogeneous catalysis *quantitatively*. First, the presence of other molecules A^* may obstruct the access to the catalytic surface. Second, certain parasite reactions happen on the boundary that implies a deactivation of the catalyst. Consequently, the reactivity K becomes dependent on the spatial position leading to an inhomogeneous boundary condition. Finally, the molecular diffusion can be applied if the mean free path of reactive molecules is lower than the geometrical features of the catalyst (in the opposite case, one deals with Knudsen diffusion [17]). Nevertheless, the simple description (1–3) permits to take into account many important features related to the catalytic process.

1.3 Electric Transport in Electrochemistry

The other example of Laplacian transport phenomena can be found in electrochemistry: the electric transport between two metallic electrodes into an electrolyte is described by the same boundary value problem. Indeed, the electric potential V obeys the Laplace equation in the bulk since the electrolyte is locally neutral. Taking one electrode of very low resistance (*counter-electrode*), one writes the corresponding boundary condition as $V = V_0$, where V_0 is the applied tension. For the other electrode of surface resistance r (*working electrode*), one obtains the mixed boundary condition by equating the volume current density $-\rho^{-1}\nabla V$ (ρ is the electrolyte resistivity) and the surface current density V/r: $\Lambda\ \partial V/\partial n = V$, where $\Lambda = r/\rho$ is again the physical length of the problem. The similar description can be brought even in the case of an alternative tension [18, 19].

For electric transport, one cannot associate directly the mixed boundary value problem with the partially reflected Brownian motion since there is no diffusing particle. From this point of view, the electrochemical problem has only a formal analogy with two previous examples. At the same time, the electrochemistry provides a wide domain for experimental studies of the influence of the irregular geometry on the (average) transport properties. Taking different metallic electrodes with micro- or macro-roughness (Fig. 2), one can measure its *spectroscopic impedance* or *admittance* corresponding to the total flux across the boundary for diffusional problems [20]. The observation of anomalous impedance behavior in [21] had provoked numerous theoretical, numerical and experimental studies of the role of a geometrical irregularity in Laplacian transport phenomena [22–32] (for more information, see [33] and references therein).

1.4 Discrete and Semi-continuous Approaches

Between different theoretical approaches developed to study the Laplacian transport phenomena, one can distinguish semi-continuous and discrete approaches: the double layer theory of Halsey and Leibig [24, 25] and the formalism of the Brownian self-transport operator introduced by Filoche and Sapoval [20] respectively. In both cases, one involves the stochastic description of particle's motion corresponding to the mixed boundary value problem (1–3). Halsey and Leibig used the Green functions in order to account the number of jump-like reflections of the Brownian motion on the boundary (see Section 4.3), while Filoche and Sapoval considered random walks on a lattice with partial reflections (see Section 4.4). Although both approaches provide a complete description of the Laplacian transport phenomena (e.g., the explicit formula for the total flux across the boundary), their essential inconvenient resides in the dependence on an artificial length scale: jump distance a for the *semi-continuous* approach of Halsey and Leibig and lattice parameter a for the *discrete* approach of Filoche and Sapoval. A physical intuition suggests that, if the description is correct, there should exist a well defined continuous limit as a tends to 0. Certain substantial arguments to justify the existence of this limit are brought in [33], but a rigorous mathematical proof is still required. In this paper, we describe a different approach which can be called *continuous*, based on the partially reflected Brownian motion. It will integrate the advantages of previous ones, being a mathematical foundation for understanding the Laplacian transport phenomena. We will return to these questions in Section 4.

2 Basic Definitions

In this section, we recall the basic definitions related to the Brownian motion and reflected Brownian motion that can be found in extensive literature, e.g., [34–40]. The familiar reader may pass over this section.

2.1 Brownian Motion and Dirichlet Boundary Value Problem

The Brownian motion can be defined in different ways [34]. Throughout this paper, we use the following definition.

Definition 2.1 *A stochastic process W_t $(t \geq 0)$ defined on the chosen probabilistic space is called one-dimensional Brownian motion (or Wiener process) started from the origin, if*

- *its trajectories are continuous almost surely (with probability 1);*
- *it starts from the origin almost surely, $\mathbb{P}\{W_0 = 0\} = 1$;*
- *its joint distribution is*

$$\mathbb{P}\{W_{t_1} \in \Gamma_1, ..., W_{t_n} \in \Gamma_n\} = \int_{\Gamma_1} dx_1 ... \int_{\Gamma_n} dx_n \, g(0, x_1\, ;\, t_1)\, g(x_1, x_2\, ;\, t_2 - t_1) \, ...\, g(x_{n-1}, x_n\, ;\, t_n - t_{n-1})$$

for any integer n, any real numbers $0 < t_1 < ... < t_n$ and arbitrary intervals Γ_1, ..., Γ_n, where $g(x, x'\, ;\, t)$ is the Gaussian density

$$g(x, x'\, ;\, t) = \frac{1}{\sqrt{2\pi t}} \exp\left[-\frac{(x - x')^2}{2t}\right] \qquad x, x' \in \mathbb{R}, \quad t \in \mathbb{R}_+ \tag{4}$$

By definition, $g(0, x\, ;\, t)dx$ is the probability to find the Brownian motion in dx vicinity of point x at time t:

$$\mathbb{P}\{W_t \in (x, x + dx)\} = g(0, x\, ;\, t)$$

The collection $W_t = (W_t^1, ..., W_t^d)$ of d independent one-dimensional Brownian motions W_t^k is called d-dimensional Brownian motion started from the origin (in the following, we will omit the pointing on the dimension). The translated stochastic process, $x + W_t$, is called Brownian motion started from the point $x \in \mathbb{R}^d$.

Various properties of the Brownian motion and its relations to other mathematical fields (like partially differential equations or potential theory) are well known and can be found in [34–40].

As one can see, the Brownian motion W_t is defined for the whole space $\mathbb{R}^d$, without any binding to a particular domain. However, the physical processes are usually confined into a certain domain $\Omega \subset \mathbb{R}^d$. The "presence" of its boundary $\partial\Omega$ can be introduced by a specific condition for a quantity we are looking for. To illustrate this notion, let us introduce the harmonic measure ω_x defined as the probability to hit the boundary $\partial\Omega$ on its different subsets for the first time.

Definition 2.2 *Let $\Omega \subset \mathbb{R}^d$ be a domain with boundary $\partial\Omega$. For any $x \in \mathbb{R}^d$, a random variable $\mathbb{T}^x = \inf\{t > 0 \; : \; (x + W_t) \in \partial\Omega\}$ is called stopping time on the boundary $\partial\Omega$ (it gives the first moment when the Brownian motion started from x hits the boundary). For any subset A from the Borel σ-algebra $\mathcal{B}(\partial\Omega)$, one defines its harmonic measure $\omega_x\{A\}$ (hitting probability) as:*

$$\omega_x\{A\} = \mathbb{P}\{W_{\mathbb{T}^x} \in A, \; \mathbb{T}^x < \infty\}$$

(we remind that the Borel σ-algebra $\mathcal{B}(\partial\Omega)$ is generated by all open subsets of $\partial\Omega$).

We gave this classical definition of the harmonic measure in order to outline that the boundary $\partial\Omega$ is present in the problem through the stopping time $\mathbb{T}^x$. In other words, its introduction does not change the definition of the Brownian motion itself. This feature considerably simplifies the following analysis.

Up to this moment, we did not specify the domain Ω and its boundary $\partial\Omega$, since the harmonic measure can be well defined for very irregular domains [41–43]. However, the following definitions will require some restrictions on the boundary. Throughout this paper, we will consider a domain $\Omega \in \mathbb{R}^d$ ($d \geq 2$) with *bounded smooth boundary* $\partial\Omega$ (twice continuous differentiable manifold). Note that this condition can be weakened in different ways, but it would require more sophisticated analysis overflowing the frames of this paper (e.g., see [59, 60]). Dealing with physical problems shown in Section 1, one can always *smooth* a given boundary $\partial\Omega$ whatever its original irregularity. Indeed, the physics provides a minimal cut-off ℓ (e.g., mean free path of diffusing or reacting molecules) which determines the "admissible" scales of the boundary. All geometrical features of the boundary smaller than ℓ should be irrelevant (otherwise, the proposed physical description would be invalid). Smoothing these geometrical elements, one obtains a boundary which is irregular on scales larger than ℓ and smooth on scales lower than ℓ. For a smooth boundary $\partial\Omega$, one can introduce the harmonic measure density $\omega_x(s)$ such that $\omega_x(s)ds$ is the probability that the Brownian motion hits the boundary in ds vicinity of the boundary point s.

The harmonic measure, generated by the Brownian motion, provides a general solution of the Dirichlet boundary value problem with a given function f on $\partial\Omega$:

$$\Delta u = 0 \quad (x \in \Omega), \qquad\qquad u = f \quad (x \in \partial\Omega) \tag{5}$$

Since the harmonic measure density is equal to the normal derivative of the Green function for the Dirichlet problem (5), one writes the solution $u(x)$ explicitly [44]:

$$u(x) = \int\limits_{\partial\Omega} f(s) \, \omega_x(s) ds$$

or as following expectation

$$u(x) = \mathbb{E}\big(f(W_{\mathbb{T}^x})\big) \tag{6}$$

One can give a physical interpretation of this mathematical relation. In order to calculate the expectation, one considers all possible trajectories of the Brownian motion started from the point $x \in \Omega$. For each trajectory terminated at boundary points s, one assigns the weight $f(s)$ and then averages over all these trajectories. Giving this interpretation, we do not discuss the mathematical realization of such average over all possible trajectories: to do this operation properly, one introduces the Wiener measure on the space of continuous functions and defines the corresponding functional integrals [34]. It is interesting to remark that this understanding traced to the Feynman integral's description of quantum mechanics [45]. Note also that the relation (6) is the mathematical foundation for Monte-Carlo numerical tools to solve the Dirichlet problem (5): launching a large number of random walkers from the point x, for each trajectory one determines its hitting point s and assigns the corresponding weight $f(s)$. The average over all random walkers gives an approximation to the value of the solution $u(x)$ at point x.

One concludes that the Brownian motion provides an efficient mathematical tool to study the Dirichlet boundary value problem. However, it becomes useless for other types of boundary conditions like, e.g., the Neumann condition. The simple physical reason is the following. As we have mentioned above, the Dirichlet boundary condition is introduced through the stopping time $\mathbb{T}^x$. It means that we are interested only in the Brownian motion W_t for times t between 0 and $\mathbb{T}^x$. Since the further motion (with $t > \mathbb{T}^x$) is completely irrelevant for this problem, one may think that the Brownian motion is *absorbed* on the boundary $\partial\Omega$ at the first hit. In other words, the Dirichlet condition corresponds to a *purely absorbing* boundary $\partial\Omega$. For the Neumann condition, the situation changes drasticly. The normal derivative representing a flux leads to the notion of *re ection* on the boundary: if one would like to fix the flux density across the boundary, certain particles should be reflected. The probabilistic description of the Neumann boundary condition necessitates thus the introduction of other stochastic process called *re ected Brownian motion* .

2.2 Re ected Brownian Motion

The fact of reflection on the boundary implies three essential distinctions with respect to the (simple) Brownian motion:

- the definition of the reflected Brownian motion will depend on the domain Ω (as a consequence, it will be more sophisticated than the definition of the Brownian motion);
- the type and direction of each reflection should be prescribed (e.g., right or oblique);
- some restrictions on the boundary $\partial\Omega$ should be introduced, e.g., the normal vector should be well defined at each point (as a consequence, the boundary cannot be very irregular).

It is not thus surprising that the definition of the reflected Brownian motion requires the study of *stochastic* differential equations. We do not pretend to reproduce the whole analysis leading to the reflected Brownian motion since one can find it in corresponding literature (e.g., see [34, 46–48]). Note that, in the case of smooth boundaries, this definition is quite classical. The situation becomes essentially more difficult when one tries to extend it for nonsmooth domains.

Definition 2.3 *Let Ω be a domain with boundary $\partial\Omega$, and $n(s)$ is a vector-valued function on $\partial\Omega$. For a given point $x \in \Omega$, one considers the stochastic equation in the following* *form:*

$$d\hat{W}_t = dW_t + n(\hat{W}_t)\mathbb{I}_{\partial\Omega}(\hat{W}_t)d\ell_t \qquad \hat{W}_0 = x, \ \ell_0 = 0 \tag{7}$$

where W_t is d-dimension Brownian motion and $\mathbb{I}_{\partial\Omega}$ is the indicator of the boundary $\partial\Omega$. By a solution of this equation, we mean a pair of almost surely continuous processes $\hat{W}_t$ and ℓ_t, satisfying (7), adapted to the underlying family of σ-fields and satisfying, with probability 1, the following conditions:

- *$\hat{W}_t$ belongs to $\Omega \cup \partial\Omega$;*
- *ℓ_t is a nondecreasing process which increases only for $t \in \mathcal{T}$, $\mathcal{T} = \{t > 0 \ : \ \hat{W}_t \in \partial\Omega\}$ having Lebesgue measure zero almost surely.*

The process $\hat{W}_t$ is called Brownian motion re ected on the boundary or re ected Brownian motion, the process ℓ_t is called local time on the boundary.

The following theorem provides the existence and uniqueness of these stochastic processes in the case of smooth boundaries.

Theorem 2.4 *Let Ω be a bounded domain with twice continuous differentiable boundary $\partial\Omega$, $n(s)$ is the vector of the inward unit normal at boundary point s (orthogonal to the boundary at s and oriented towards the domain). For a given point $x \in \Omega \cup \partial\Omega$, the stochastic equation (7) possesses a unique solution, i.e., there exist the re ected Brownian motion $\hat{W}_t$ and the local time on the boundary ℓ_t satisfying the above conditions, and they are unique.*

Proof can be found in [34, 46].

We should note that this theorem can be extended in different ways. For example, one can consider the Brownian motion, reflected on the boundary in the direction given by another vector-valued field than the field $n(s)$ of the inward unit normals. The assumption that the domain is bounded can be replaced by a more subtle hypothesis that allows to extend the definition of the reflected Brownian motion for some classes of unbounded domains. At last, one may define this motion for a general case of second order elliptic differential operators (with certain restrictions on their coefficients). The interested reader may consult the corresponding literature, e.g., [34, 38].

Although the rigorous mathematical definition of stochastic differential equations is more difficult than in the case of ordinary differential equations, an intuitive meaning of its elements remains qualitatively the same. For example, the stochastic equation (7) states that an infinitesimal variation $d\hat{W}_t$ of the reflected Brownian motion $\hat{W}_t$ in the domain Ω is governed only by the variation dW_t of the (simple) Brownian motion W_t (the second term vanishes due to the indicator $\mathbb{I}_{\partial\Omega}$). When the motion hits the boundary, the second term does not allow to leave the domain leading to a variation directed along the inward unit normal $n(s)$ towards the interior of the domain. On the other hand, each hit of the boundary increases the local time ℓ_t. Consequently, the single stochastic equation (7) defines simultaneously two random processes, $\hat{W}_t$ and ℓ_t, strongly dependent each of other.

As an example, one can consider the one-dimensional Brownian motion reflected at zero which can be written as mirror reflection of the (simple) Brownian motion: $\hat{W}_t = |x + W_t|$. Applying Itô's formula to this function, one obtains:

$$\hat{W}_t = |x + W_t| = x + \int_0^t \mathrm{sign}(x + W_{t'})dW_{t'} + \frac{1}{2}\int_0^t \delta(x + W_{t'})dt'$$

One can show that the second term corresponds to a Brownian motion W'_t, whereas the third term, denoted as ℓ_t, is a continuous, nondecreasing random process which increases only on the set $\mathcal{T} = \{t > 0 \ : \ x + W_t = 0\}$ of the Lebesgue measure zero. The previous expression can be thus written as $\hat{W}_t = x + W'_t + \ell_t$ or, in differential form, as $d\hat{W}_t = dW'_t + d\ell_t$ which is the particular case of the stochastic equation (7). For the local time ℓ_t, Lévy proved the following representation [36, 37]:

$$\ell_t = \lim_{a \to 0} \frac{1}{2a} \int_0^t \mathbb{I}_{[0,a]}(\hat{W}_{t'})dt' \tag{8}$$

This relation makes explicit the meaning of the local time ℓ_t: it shows how "many times" the reflected Brownian motion passed in an infinitesimal vicinity of zero up to the moment t. We give also the other useful representation for the local time (proposed also by Lévy):

$$\ell_t = \lim_{a \to 0} a\mathcal{N}_t(a) \tag{9}$$

where $\mathcal{N}_t(a)$ is the number of passages of the reflected Brownian motion through the interval $[0, a]$ up to the moment t. If one introduces a sequence of stopping times at points 0 and a,

$$\tau_0^{(0)} = \inf\{t > 0 \ : \ \hat{W}_t = 0\} \qquad \tau_0^{(a)} = \inf\{t > \tau_0^{(a)} \ : \ \hat{W}_t = a\}$$
$$\tau_n^{(0)} = \inf\{t > \tau_{n-1}^{(a)} \ : \ \hat{W}_t = 0\} \qquad \tau_n^{(a)} = \inf\{t > \tau_{n-1}^{(0)} \ : \ \hat{W}_t = a\}$$

the number of passages can be defined as

$$\mathcal{N}_t(a) = \sup\{n > 0 \ : \ \tau_n^{(0)} < t\}$$

Note that these representations can be extended for a general case of d-dimensional reflected Brownian motion.

3 Partially Re ected Brownian Motion

3.1 Definition and Certain Properties

Bearing in mind the description of the Laplacian transport phenomena, we would like to extend the reflected Brownian motion in order to be able to deal with the mixed boundary condition (3).

Definition 3.1 *For a given domain $\Omega \subset \mathbb{R}^d$ with smooth bounded boundary $\partial\Omega$, let $\hat{W}_t$ be the re ected Brownian motion started from $x \in \Omega$ and ℓ_t be the related local time process. Let χ be the random variable, independent of $\hat{W}_t$ and ℓ_t and distributed according to the exponential law with a positive parameter Λ:*

$$\mathbb{P}\{\chi \geq \lambda\} = \exp[-\lambda/\Lambda] \qquad (\lambda \geq 0) \tag{10}$$

The stopping time

$$\mathbb{T}^x_\Lambda = \inf\{t > 0 \ : \ \ell_t \geq \chi\}$$

shows the first moment when the local time process ℓ_t exceeds the random variable χ. The process $\hat{W}_t$ conditioned to stop at random moment $t = \mathbb{T}^x_\Lambda$ is called partially re ected Brownian motion (PRBM).

First of all, we stress that the partially reflected Brownian motion is not a *new* stochastic process: it reproduces completely the reflected Brownian motion $\hat{W}_t$ up to the moment $\mathbb{T}^x_\Lambda$. The only difference between them resides in the fact that we are not interested in what happens after this moment. Consequently, the condition to stop at $t = \mathbb{T}^x_\Lambda$ may be thought as an absorption by the boundary $\partial\Omega$. It explains the term "partially reflected": after multiple reflections, the process will be absorbed on the boundary (see Section 4 for further comments). Roughly speaking, the whole term "partially reflected Brownian motion" is a shorter version of the phrase "reflected Brownian motion conditioned to stop at random moment $\mathbb{T}^x_\Lambda$".

In the particular case $\Lambda = 0$, the exponential distribution (10) degenerates to $\mathbb{P}\{\chi = 0\} = 1$ and $\mathbb{P}\{\chi > 0\} = 0$. Consequently, the stopping time becomes: $\mathbb{T}^x_0 = \inf\{t > 0 \ : \ \ell_t > 0\}$. Since the first moment of an increase of the local time process ℓ_t corresponds to the first hit of the boundary $\partial\Omega$, one obtains the stopping time of the (simple) Brownian motion: $\mathbb{T}^x_0 = \mathbb{T}^x$. One concludes that, for $\Lambda = 0$, the partially reflected Brownian motion becomes the Brownian motion conditioned to stop at the first hit of the boundary.

The partially reflected Brownian motion allows to introduce a measure to quantify the absorption on different subsets of the boundary $\partial\Omega$.

Definition 3.2 *For any subset A from the Borel σ-algebra $\mathcal{B}(\partial\Omega)$, one defines its spread harmonic measure $\omega_{x,\Lambda}\{A\}$ as:*

$$\omega_{x,\Lambda}\{A\} = \mathbb{P}\{\hat{W}_{\mathbb{T}^x_\Lambda} \in A,\ \mathbb{T}^x_\Lambda < \infty\}$$

As for the harmonic measure, $\omega_{x,\Lambda}\{A\}$ satisfies the properties of a probabilistic measure, in particular, $\omega_{x,\Lambda}\{\partial\Omega\} = 1$. When Λ goes to 0, the spread harmonic measure tends to the harmonic measure: $\omega_{x,\Lambda}\{A\} \to \omega_x\{A\}$.

Since the definition of the PRBM requires the smoothness of the boundary, one can introduce the spread harmonic measure density $\omega_{x,\Lambda}(s)$.

Dealing with the Brownian motion, one could formally take the starting point x on the boundary $\partial\Omega$, but it would lead to trivial results: the stopping time $\mathbb{T}^x$ becomes 0 and the harmonic measure ω_x is degenerated to the Dirac point measure: $\omega_x\{A\} = \mathbb{I}_A(x)$ (if $x \in \partial\Omega$). In the case of the partially reflected Brownian motion, the starting point x can belong to the domain Ω or to its boundary $\partial\Omega$: in both cases the spread harmonic measure has nontrivial properties.

It is convenient to separate each random trajectory of the PRBM in two parts, before and after the first hit of the boundary. The first part, $\hat{W}_{0\leq t\leq \mathbb{T}^x}$, coincides with the (simple) Brownian motion started from x and conditioned to stop on the boundary, while the second part, $\hat{W}_{\mathbb{T}^x\leq t\leq \mathbb{T}^x_\Lambda}$, coincides with the reflected Brownian motion started on the boundary (at the first hitting point) and conditioned to stop on the same boundary at random moment $\mathbb{T}^x_\Lambda$. Since these two parts are independent, one can write the spread harmonic measure density as

$$\omega_{x,\Lambda}(s) = \int\limits_{\partial\Omega} ds' \omega_x(s')\, T_\Lambda(s', s) \qquad\qquad T_\Lambda(s', s) \equiv \omega_{s',\Lambda}(s) \tag{11}$$

The integral kernel $T_\Lambda(s, s')$ represents the probability density that the PRBM started from the boundary point s' is stopped (absorbed) in an infinitesimal vicinity of the boundary point s. Consequently, it is sufficient to determine the probabilities of displacements between two boundary points in order to reconstruct the whole spread harmonic measure density.

Lemma 3.3 *For any subset A from $\mathcal{B}(\partial\Omega)$ and fixed positive Λ, the spread harmonic measure $\omega_{x,\Lambda}\{A\}$, considered as function of x, solves the mixed boundary value problem:*

$$\Delta\omega_{x,\Lambda}\{A\} = 0 \quad (x \in \Omega), \qquad\qquad \left[I - \Lambda\frac{\partial}{\partial n}\right]\omega_{x,\Lambda}\{A\} = \mathbb{I}_A(x) \quad (x \in \partial\Omega) \tag{12}$$

This lemma generalizes the Kakutani theorem for the harmonic measure (when $\Lambda = 0$) [49]. We do not reproduce the proof of this lemma since it would require many technical details. It can be also written for the spread harmonic measure density:

Lemma 3.4 *For any boundary point $s \in \partial\Omega$ and fixed positive Λ, the spread harmonic measure density $\omega_{x,\Lambda}(s)$, considered as function of x, satisfies the following conditions:*

$$\Delta\omega_{x,\Lambda}(s) = 0 \quad (x \in \Omega), \qquad\qquad \left[I - \Lambda\frac{\partial}{\partial n}\right]\omega_{x,\Lambda}(s) = \delta(s - x) \quad (x \in \partial\Omega) \tag{13}$$

where $\delta(s - x)$ is the Dirac function (distribution) on the boundary.

According to this lemma, one can write the solution of a general mixed boundary value problem with a given function f on $\partial\Omega$ and fixed positive Λ,

$$\Delta u = 0 \quad (x \in \Omega), \qquad\qquad \left[I - \Lambda\frac{\partial}{\partial n}\right] u = f \quad (x \in \partial\Omega)$$

in two equivalent forms:

$$u(x) = \int\limits_{\partial\Omega} f(s)\, \omega_{x,\Lambda}(s) ds = \mathbb{E}\big(f(\hat{W}_{\mathbb{T}^x_\Lambda})\big)$$

Again, one can give a physical interpretation of this relation: one averages the function f over all possible trajectories of the partially reflected Brownian motion started from the point x. Each trajectory is weighted by $f(s)$ according to the boundary point s of its final absorption.

3.2 Planar Surface

We remind that the physical motivation of this study is a possibility to consider the partially reflected Brownian motion as the random trajectory of diffusing particles towards semi-permeable interfaces. Indeed, the boundary value problem (1–3) is an averaged description for the particle's concentration, while the stochastic description permits to "follow" the trajectory of each particle providing new information: typical or average distance between the first hitting point and the final absorption point; proportion of "flatten" trajectories, going near the interface, with respect to remote trajectories, moving away from the interface and then returning to it, etc. In this subsection, we briefly consider the particular case of the planar surface (boundary of a half space), when the partially reflected Brownian motion can be constructed in a simple way, without stochastic equations. Consequently, many related characteristics can be obtained explicitly. In addition, this construction for the half space provides an example of the PRBM for an unbounded domain.

Let Ω be the upper half space, $\Omega = \{x \in \mathbb{R}^d \ : \ x_d > 0\}$, with smooth boundary $\partial\Omega = \{x \in \mathbb{R}^d \ : \ x_d = 0\}$. Let W_t^k are d independent Brownian motions started from the origin. Then, the Brownian motion, started from a given point $x \in \Omega$ and reflected on the boundary $\partial\Omega$, can be written in a simple way as $(x_1 + W_t^1, \ ..., \ x_{d-1} + W_t^{d-1}, |x_d + W_t^d|)$. The particular simplification is brought by the fact that reflections happen in a single direction, being involved through the one-dimensional reflected Brownian motion $|x_d + W_t^d|$. Without loss of generality, we can consider the reflected Brownian motion started from the origin ($x = 0$). The translational invariance along the hyperplane $\partial\Omega$ permits to move the starting point in $\partial\Omega$, whereas the convolution property (11) allows displacements in orthogonal direction. The local time process ℓ_t can be introduced either through the stochastic equations (7), or with the help of Lévy's formulae (8) or (9).

Lemma 3.5 *Let Ω be the upper half space, $\Omega = \{x \in \mathbb{R}^d \ : \ x_d > 0\}$. For any positive Λ, the stopping time $\mathbb{T}_\Lambda^0$, defined in 3.1, is distributed according to*

$$\mathbb{P}\{\mathbb{T}_\Lambda^0 \in (t, t+dt)\} = \rho_\Lambda(t)dt, \qquad \rho_\Lambda(t) = \int\limits_0^\infty \frac{z\, e^{-z^2/2t} e^{-z/\Lambda}}{\Lambda\sqrt{2\pi}\, t^{3/2}}\, dz \tag{14}$$

Proof. Since the local time ℓ_t and the random variable χ are independent, one can write the probability $\mathbb{P}\{\mathbb{T}_\Lambda^0 \in (t, t+dt)\}$ as

$$\mathbb{P}\{\ \mathbb{T}_\Lambda^0 \in (t, t+dt)\ \} = \int\limits_0^\infty \mathbb{P}\Big\{\ \inf\{\tau > 0 \ : \ \ell_\tau = z\} \in (t, t+dt)\ \Big\}\ \mathbb{P}\{\ \chi \in (z, z+dz)\ \}$$

Then, the first factor is the well known density of the inverse local time process [35],

$$\mathbb{P}\Big\{\ \inf\{\tau > 0 \ : \ \ell_\tau = z\} \in (t, t+dt)\ \Big\} = dt\, \frac{z\, e^{-z^2/2t}}{\sqrt{2\pi}\, t^{3/2}}$$

whereas the second factor is given by the exponential law density (10) that implies (14). □

Note that the integral in (14) can be represented with the help of the Gaussian error function

$$\rho_\Lambda(t) = \frac{1}{2\Lambda^2}\left[\frac{1}{\sqrt{\pi}}\frac{1}{\sqrt{t/2\Lambda^2}} - \mathcal{K}\left(\sqrt{t/2\Lambda^2}\right)\right], \qquad \mathcal{K}(z) = \frac{2}{\sqrt{\pi}}\int\limits_z^\infty e^{z^2-x^2}dx$$

One finds the asymptotic behavior of the density $\rho_\Lambda(t)$:

$$\rho_\Lambda(t) \sim \left(\sqrt{2\pi}\Lambda\right)^{-1} t^{-1/2} \quad (t\to 0), \qquad \rho_\Lambda(t) \sim \left(\sqrt{2\pi}/\Lambda\right)^{-1} t^{-3/2} \quad (t\to\infty)$$

Once the distribution of stopping time $\mathbb{T}^0_\Lambda$ is determined, one can calculate the spread harmonic measure density $\omega_{x,\Lambda}(s)$.

Lemma 3.6 *Let Ω be the upper half space, $\Omega = \{x \in \mathbb{R}^d \; : \; x_d > 0\}$. For any positive Λ, the spread harmonic measure density $\omega_{x,\Lambda}(s)$ is*

$$\omega_{x,\Lambda}(s_1,...,s_{d-1}) = \int\limits_{-\infty}^{\infty}\cdots\int\limits_{-\infty}^{\infty}\frac{dk_1...dk_{d-1}}{(2\pi)^{d-1}}\exp\left[-i\sum_{j=1}^{d-1}k_j(x_j-s_j)\right]\frac{e^{-x_d|k|}}{1+\Lambda|k|} \qquad (15)$$

where $|k| = \sqrt{k_1^2+...+k_{d-1}^2}$.

Proof. First, the probability kernel $T_\Lambda(s,s')$, defined for two boundary points $s, s' \in \partial\Omega$, is translationally invariant in the hyperplane $\partial\Omega$, $T_\Lambda(s,s') = t_\Lambda(s-s')$, where

$$t_\Lambda(s_1,...,s_{d-1})ds_1...ds_{d-1} \equiv \mathbb{P}\left\{W^1_{\mathbb{T}^0_\Lambda} \in (s_1, s_1+ds_1), \; ..., \; W^{d-1}_{\mathbb{T}^0_\Lambda} \in (s_{d-1}, s_{d-1}+ds_{d-1})\right\}$$

The stopping time $\mathbb{T}^0_\Lambda$ is related to the orthogonal motion and, consequently, independent of lateral motions W^1_t, ..., W^{d-1}_t. Therefore, the previous probability can be written as

$$t_\Lambda(s_1,...,s_{d-1})ds_1...ds_{d-1} =$$

$$= \int\limits_0^\infty \mathbb{P}\left\{W^1_t \in (s_1, s_1+ds_1), \; ..., \; W^{d-1}_t \in (s_{d-1}, s_{d-1}+ds_{d-1})\right\}\rho_\Lambda(t)dt$$

Since the lateral motions are independent between themselves, the first factor is equal to the product of Gaussian densities (4):

$$t_\Lambda(s_1,...,s_{d-1}) = \int\limits_0^\infty dt\; \rho_\Lambda(t)\prod_{j=1}^{d-1}\frac{e^{-s_j^2/2t}}{\sqrt{2\pi t}}$$

Using the integral representation (14), one finds

$$t_\Lambda(s_1,...,s_{d-1}) = \frac{\Gamma(d/2)}{\pi^{d/2}\,\Lambda}\int\limits_0^\infty dz\,\frac{z\,e^{-z/\Lambda}}{\left[s_1^2+...+s_{d-1}^2+z^2\right]^{d/2}}$$

where $\Gamma(z)$ stands for Euler gamma function (see [33] for details).

Substituting the well known harmonic measure density $\omega_x(s)$ for the upper half space (generalized Cauchy distribution),

$$\omega_x(s_1, ..., s_{d-1}) = \frac{\Gamma(d/2)}{\pi^{d/2}} \frac{x_d}{\left[(x_1 - s_1)^2 + ... + (x_{d-1} - s_{d-1})^2 + (x_d)^2\right]^{d/2}}$$

into convolution (11), one finally obtains the expression (15) for the spread harmonic measure density. □

One can easily verify that the spread harmonic measure density $\omega_{x,\Lambda}(s)$ and the probability kernel $T_\Lambda(s, s')$ satisfy the following conditions:

1. Normalization condition:

$$\int\limits_{\partial\Omega} ds\, \omega_{x,\Lambda}(s) = 1 \qquad\qquad \int\limits_{\partial\Omega} ds'\, T_\Lambda(s, s') = 1$$

2. Dirichlet limit ($\Lambda \to 0$):

$$\omega_{x,\Lambda}(s) \longrightarrow \omega_x(s) \qquad\qquad T_\Lambda(s, s') \longrightarrow \delta(s - s')$$

3. Translational invariance:

$$\omega_{x,\Lambda}(s) = \omega_{x-s,\Lambda}(0) \qquad\qquad T_\Lambda(s, s') = T_\Lambda(s - s', 0) \equiv t_\Lambda(s - s')$$

One can also deduce the asymptotic behavior of the function $t_\Lambda(s)$ as $|s| \to 0$ or $|s| \to \infty$. For these purposes, it is convenient to define the new function $\eta_d(z)$ by relation

$$t_\Lambda(s) = \eta_d\big(|s|/\Lambda\big)\, \omega_{(0,...,0,\Lambda)}(s)$$

where the second term is the harmonic measure density for the Brownian motion started from the point $(\underbrace{0, ..., 0}_{d-1}, \Lambda)$. Using the explicit formulae for $t_\Lambda(s)$ and $\omega_{(0,...,0,\Lambda)}(s)$, one obtains:

$$\eta_d(z) = \left(1 + z^2\right)^{d/2} \int\limits_0^\infty \frac{t\, e^{-t}\, dt}{(t^2 + z^2)^{d/2}} \qquad (16)$$

Its asymptotic behavior for z going to infinity is

$$\eta_d(z) = 1 - \frac{5d}{2}\, z^{-2} + O(z^{-4}) \qquad (17)$$

whereas for z going to 0, one has

$$\eta_d(z) \sim \frac{\Gamma(d/2)}{\pi^{d/2}}\, z^{2-d} \quad (d > 2), \qquad\qquad \eta_d(z) \sim \frac{1}{\pi} \ln z \quad (d = 2)$$

These relations can be used for qualitative study of the partially reflected Brownian motion. For instance, one identifies the parameter Λ as a characteristic length scale of the problem: the magnitude of any distance (e.g., $|s|$) has to be compared with Λ. Interestingly, the asymptotic behavior (17) for large z means that the function $t_\Lambda(s)$ is close to the harmonic measure density $\omega_{(0,...,0,\Lambda)}(s)$. Roughly speaking, for large $|s|/\Lambda$, the partially reflected Brownian motion started from the origin is qualitatively equivalent to the (simple) Brownian motion started from the point $(0,...,0,\Lambda)$. In other words, the partial reflections on the boundary leads to a *spreading* of the harmonic measure with characteristic scale Λ (see also relation (11)).

The knowledge of the probability kernel $T_\Lambda(s,s')$ provides an important information about the partially reflected Brownian motion in the (upper) half space. As an example, we calculate the probability $\mathcal{P}_\Lambda(r)$ that the PRBM started from the origin is finally absorbed on the disk $B_r^{d-1} = \{ (x_1,...,x_d) \in \mathbb{R}^d \ : \ x_1^2 + ... + x_{d-1}^2 \leq r^2, \ x_d = 0 \}$ of radius r centered at the origin:

$$\mathcal{P}_\Lambda(r) = \int\limits_{B_r^{d-1}} ds\, t_\Lambda(s) = \frac{2\,\Gamma(\frac{d}{2})}{\Gamma(\frac{d-1}{2})\sqrt{\pi}} \int\limits_0^\infty te^{-t} dt \int\limits_0^{r/\Lambda} \frac{x^{d-2}\, dx}{[x^2+t^2]^{d/2}} \tag{18}$$

This probability shows how far the partially reflected Brownian motion can go away after the first hit of the boundary. One sees that this function depends only on the ratio r/Λ, going to 0 for small radii and to 1 for large radii. Again, the parameter Λ is the characteristic length scale of the problem. In two-dimensional case, $\mathcal{P}_\Lambda(\Lambda/2)$ is the probability that the PRBM is absorbed on the linear segment of length Λ, centered at the origin (the first hitting point). The numerical calculation gives $\mathcal{P}_\Lambda(\Lambda/2) \simeq 0.4521$, i.e., about half of the particles is absorbed on this region. In other words, the length of the characteristic absorption region (where half of the particles is absorbed) is equal approximately to Λ. It has been shown recently that this result is qualitatively valid for certain irregular boundaries [50]. Roughly speaking, if the one-dimensional boundary (curve) has no deep pores (fjords) and its perimeter is large with respect to the scale Λ, then the curvilinear interval of length Λ, centered on the first hitting point, absorbs approximately half of the diffusing particles.

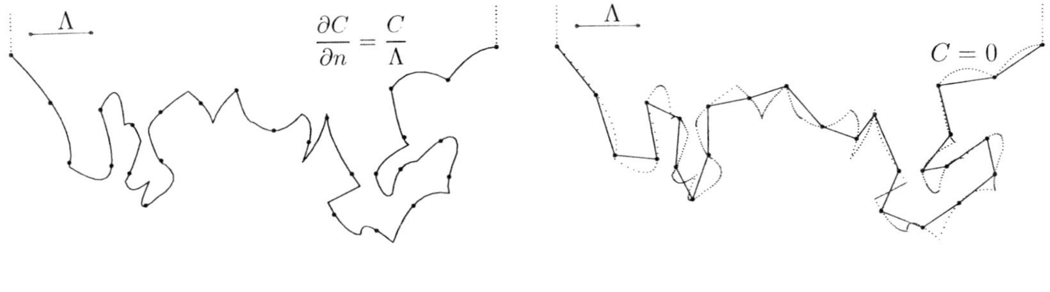

Figure 3: Land Surveyor Approximation: the total flux across the boundary can be approximately calculated when the mixed boundary condition $\partial C/\partial n = C/\Lambda$ on a given irregular curve (on the left) is replaced by the Dirichlet condition $C = 0$ on the coarse-grained boundary (on the right). The last one is obtained by replacing curvilinear intervals of length Λ by corresponding linear chords.

This result can be considered as a first mathematical justification of the *Land Surveyor Approximation* (LSA) developed by Sapoval [18]. According to this approximation, one can coarse-grain a given one-dimensional interface (curve) with physical scale Λ in order to replace the mixed boundary condition $[I - \Lambda\partial/\partial n]C = 0$ by the Dirichlet condition $C = 0$ (see Fig. 3). Some heuristic physical arguments allow to state that the total flux across the irregular semi-permeable interface is equal approximately to the total flux across this coarse-grained boundary with Dirichlet condition. This statement provides a simple but powerful tool to study the Laplacian transport phenomena. The land surveyor approximation had been checked numerically [29, 30], but not mathematically. The study of the partially reflected Brownian motion provides its justification and further understanding. Actually, the coarse-grain procedure generates the regions of length Λ, where about half of the particles is absorbed. The LSA is based on two simplifications which can be clearly explained in terms of diffusing particles:

1. All particles arrived to the characteristic absorption region are absorbed (Dirichlet condition). This approximation does not take into account half of the particles which escapes this region.

2. The deterministic regions, generated by the coarse-grain procedure, cannot be centered on a random position of the first hit of the boundary.

Although these simplifications seem to be rough, the numerical simulations showed that the LSA reproduces the transport properties with relatively good accuracy. However, this approximation has no small parameter which would allow to control its applicability. More accurate approaches will be discussed below (see Section 4).

One can go further by extending the land surveyor approximation to the three-dimensional case, which remains still not well studied. Indeed, the numerical calculation leads to $\mathcal{P}_\Lambda(\Lambda) \simeq 0.4611$, i.e., about half of the particles is absorbed on the disk of radius Λ centered on the first hitting point. Consequently, if one finds a convenient cover of a semi-permeable irregular interface by disk-like sets of characteristic radius Λ, the LSA may be still valid, i.e., the total flux across a given interface can be approximated by the total flux across the perfectly absorbing coarse-grained interface (with Dirichlet condition). An accurate mathematical formulation of this extension and its numerical verification present still open problems.

3.3 Relation to the Dirichlet-to-Neumann Operator

The construction of the partially reflected Brownian motion for a given domain Ω requires the resolution of the stochastic differential equation (7), a quite difficult problem. Fortunately, many characteristics of this process, e.g. the spread harmonic measure density $\omega_{x,\Lambda}(s)$, can be obtained in another way. This subsection is devoted to the Dirichlet-to-Neumann operator and its relation to the partially reflected Brownian motion.

Definition 3.7 *For a given domain $\Omega \subset \mathbb{R}^d$ ($d \geq 2$) with smooth bounded boundary $\partial\Omega$, let $u : \Omega \cup \partial\Omega \to \mathbb{R}$ be a harmonic function with Dirichlet condition $u = f$, a function f being from the Sobolev space $H^1(\partial\Omega)$ (in other words, u is the the solution of the boundary value problem (5)). Applying the normal derivative to u, one obtains a new function $g = \partial u/\partial n$*

belonging to the square integrable functions space $L^2(\partial\Omega)$. Then the operator $\mathcal{M}$, acting from $H^1(\partial\Omega)$ to $L^2(\partial\Omega)$, which associates the new function g with a given function f, is called Dirichlet-to-Neumann operator.

It is known that the Dirichlet-to-Neumann operator $\mathcal{M}$ is self-adjoint pseudodifferential operator of the first order, with discrete positive spectrum $\{\mu_\alpha\}$ and smooth eigenfunctions forming a complete basis in $L^2(\partial\Omega)$ [51–60]. For this operator, one can define its resolvent operator $T_\Lambda = [I+\Lambda\mathcal{M}]^{-1}$, called *spreading operator*. This is an analytic operator function in the whole complex plane except a denumerable set of points, $\mathbb{C}\backslash\{-\mu_\alpha^{-1}\}$. In particular, T_Λ is well defined for any positive Λ.

Lemma 3.8 *For any strictly positive Λ, the spreading operator T_Λ acts from $L^2(\partial\Omega)$ to $L^2(\partial\Omega)$ as a compact integral operator,*

$$[T_\Lambda f](s) = \int\limits_{\partial\Omega} ds'\ f(s')\ T_\Lambda(s', s)$$

where the kernel $T_\Lambda(s, s')$ is given by (11).

Proof. The probability kernel $T_\Lambda(s, s')$ is a positive function satisfying the normalization:

$$\int\limits_{\partial\Omega} T_\Lambda(s, s')ds' = 1$$

since the partially reflected Brownian motion is conditioned to be finally absorbed on the boundary. Therefore, one obtains:

$$\int\limits_{\partial\Omega}\int\limits_{\partial\Omega} ds\ ds'\ |T_\Lambda(s, s')|^2 = S_{tot} < \infty$$

where S_{tot} is the total surface area of the boundary $\partial\Omega$. The integral operator T_Λ defined by the kernel $T_\Lambda(s, s')$ is a Hilbert-Schmidt operator and, consequently, a compact operator.

The boundary condition in lemma 3.4 can be written with the help of the Dirichlet-to-Neumann operator:

$$[I + \Lambda\mathcal{M}]T_\Lambda(s, s') = \delta(s - s')$$

that implies that the integral operator T_Λ coincides with the resolvent $[I + \Lambda\mathcal{M}]^{-1}$ of the Dirichlet-to-Neumann operator $\mathcal{M}$. □

This simple lemma creates a "bridge" between the partially reflected Brownian motion and the Dirichlet-to-Neumann operator. In particular, the relation (11) for the spread harmonic measure density $\omega_{x,\Lambda}(s)$ can be now understood as application of the spreading operator T_Λ to the harmonic measure density $\omega_x(s)$. Consequently, once the Dirichlet-to-Neumann operator is constructed for a given domain, one can calculate the density $\omega_{x,\Lambda}(s)$ without solving the stochastic differential equations (7).

The self-adjointness of the Dirichlet-to-Neumann operator allows to involve efficient tools of the spectral theory. For example, one can rewrite the relation (11) as spectral decomposition of the harmonic measure density on eigenfunctions $\mathbf{V}_\alpha$ of the operator $\mathcal{M}$:

$$\omega_{x,\Lambda}(s) = \sum_\alpha \frac{(\omega_x \cdot \mathbf{V}^*_\alpha)_{L^2}}{1+\Lambda\mu_\alpha} V_\alpha(s) \qquad (\omega_x \cdot \mathbf{V}^*_\alpha)_{L^2} = \int\limits_{\partial\Omega} \omega_x(s')\, \mathbf{V}^*_\alpha(s')\, ds' \tag{19}$$

where $(\,\cdot\,)_{L^2}$ denotes the scalar product in $L^2(\partial\Omega)$ space. The advantage of this relation is an explicit dependence on the physical parameter Λ.

3.4 Examples

In order to illustrate the underlying concepts, we consider several examples.

3.4.1 Two-Dimensional Disk

We are going to study the partially reflecting Brownian motion in a unit disk, $\Omega = \{\, x \in \mathbb{R}^2 \,:\, |x| < 1 \,\}$ (its boundary is a unit circle, $\partial\Omega = \{\, x \in \mathbb{R}^2 \,:\, |x| = 1 \,\}$).

In this case, the harmonic measure density $\omega_x(s) \equiv \omega(r,\theta)$ is a function of two real variables: the distance $0 \le r < 1$ between the starting point $x \in \Omega$ and the origin, and the angle $0 \le \theta < 2\pi$ between directions onto points x and $s \in \partial\Omega$ from the origin. The harmonic measure density is known as Poisson kernel:

$$\omega(r,\theta) = \frac{1-r^2}{2\pi(1-2r\cos\theta + r^2)} \tag{20}$$

The rotational invariance of the domain Ω implies that the eigenbasis of the Dirichlet-to-Neumann operator $\mathcal{M}$ is the Fourier basis,

$$\mathbf{V}_\alpha(\theta) = \frac{e^{i\alpha\theta}}{\sqrt{2\pi}} \qquad (\alpha \in \mathbb{Z})$$

Taking a Fourier harmonics as boundary condition, $u(r=1,\theta) = e^{i\alpha\theta}$, one finds the regular solution of the corresponding Dirichlet problem, $u(r,\theta) = r^{|\alpha|}e^{i\alpha\theta}$. Since the normal derivative coincides with the radius derivative, one obtains the eigenvalues of the Dirichlet-to-Neumann operator:

$$\mu_\alpha = |\alpha| \qquad (\alpha \in \mathbb{Z})$$

These eigenvalues are doubly degenerated (expect $\mu_0 = 0$).

The spread harmonic measure density is given by relation (19):

$$\omega_{x,\Lambda}(s) \equiv \omega_\Lambda(r,\theta) = \frac{1}{2\pi} \sum_{\alpha=-\infty}^{\infty} \frac{r^{|\alpha|}\, e^{i\alpha\theta}}{1+\Lambda|\alpha|}$$

(the scalar product of the harmonic measure density $\omega_x(s)$ and eigenfunctions $\mathbf{V}_\alpha(\theta)$ is shown to be equal to $r^{|\alpha|}$, with $r = |x|$). In the case $\Lambda = 0$, one finds the Poisson representation for the harmonic measure density (20) just as required. The kernel of the resolvent operator T_Λ is

$$T_\Lambda(\theta, \theta') = \frac{1}{2\pi} \sum_{\alpha=-\infty}^{\infty} \frac{e^{i\alpha(\theta-\theta')}}{1 + \Lambda|\alpha|}$$

For the exterior problem, when $\Omega = \{x \in \mathbb{R}^2 \; : \; |x| > 1\}$, one obtains exactly the same results.

3.4.2 Three-Dimensional Ball

The similar arguments can be applied for higher dimensions. For example, in the three-dimensional case, one considers the unit ball $\Omega = \{ \; x \in \mathbb{R}^3 \; : \; |x| < 1 \; \}$. The harmonic measure density is known to be

$$\omega_x(s) \equiv \omega(r, \theta) \equiv \frac{1 - r^2}{4\pi\left(1 - 2r\cos\theta + r^2\right)^{3/2}} \qquad (s \in \partial\Omega)$$

where $r = |x| < 1$ is the distance between the starting point $x \in \Omega$ and the origin, and θ is the angle between directions onto points x and $s \in \partial\Omega$ from the origin. The development of this function on spherical harmonics is

$$\omega_x(s) = \sum_{l=0}^{\infty} \sum_{m=-l}^{l} r^l \, Y_{l,m}(s) \, Y_{l,m}(x/r)$$

The rotational symmetry of the problem implies that the eigenbasis of the Dirichlet-to-Neumann operator is formed by spherical harmonics $Y_{l,m}$. A regular solution of the Dirichlet problem (5) in the unit ball can be written in spherical coordinates r, θ and φ as

$$u(r, \theta, \varphi) = \sum_{l=0}^{\infty} \sum_{m=-l}^{l} f_{l,m} \, r^l \, Y_{l,m}(\theta, \varphi)$$

where $f_{l,m}$ are coefficients of the development of a given function f (Dirichlet condition) on the complete basis of spherical harmonics. Since the normal derivative coincides with the radius derivative, one obtains

$$[\mathcal{M}f](\theta, \varphi) = \left(\frac{\partial u}{\partial n}\right)_{\partial\Omega} = \left(\frac{\partial u}{\partial r}\right)_{r=1} = \sum_{l=0}^{\infty} \sum_{m=-l}^{l} f_{l,m} \, l \, Y_{l,m}(\theta, \varphi)$$

i.e., the eigenvalues of the Dirichlet-to-Neumann operator $\mathcal{M}$ are

$$\mu_l = l \qquad (l \in \{0, 1, 2, ...\}) \tag{21}$$

Note that the l-th eigenvalue is degenerated $n_l = (2l + 1)$ times.

Interestingly[2], the Dirichlet-to-Neumann operator $\mathcal{M}$ for the unit ball $\Omega = \{x \in \mathbb{R}^3 \; : \; |x| < 1\}$ coincides with an operator introduced by Dirac in quantum mechanics [61]. It is known that the hydrogen atom is described by three quantum numbers: the main

[2]The author thanks S. Shadchin for valuable discussions on this relation.

quantum number n, the orbital quantum number l and magnetic quantum number m. Two last numbers are associated with indices of spherical harmonics. Thus, the Dirichlet-to-Neumann operator for the ball is apparently the orbital quantum number operator for the hydrogen atom. In particular, the degeneracy of eigenvalues of this operator can be understood from the point of view of spin degeneracy.

The spread harmonic measure density $\omega_{x,\Lambda}(s)$ and the spreading operator kernel $T_\Lambda(s, s')$ can be written explicitly as spectral decompositions on the eigenbasis of the Dirichlet-to-Neumann operator $\mathcal{M}$ as in the two-dimensional case.

The eigenvalues of the Dirichlet-to-Neumann operator for d-dimensional unit ball are still given by (21) with degeneracy

$$n_l = \frac{(2l+d-2)}{(d-2)} \frac{(l+d-3)!}{(d-3)!\,l!}$$

The exterior problem for $\Omega = \{x \in \mathbb{R}^3 \ : \ |x| > 1\}$ can be considered in the same manner. The harmonic measure density is given as $\tilde{\omega}_x(s) = (1/r)\ \omega(1/r, \theta)$, with $r = |x| > 1$. Using the same development on spherical harmonics, one obtains $\mu_l = l + 1$ with $l \in \{0, 1, 2, ...\}$. In particular, the lowest eigenvalue $\mu_0 = 1$ is strictly positive. This difference with respect to the spectrum μ_l for the interior problem has a simple probabilistic origin: the Brownian motion in three dimensions is transient, i.e., there is a positive probability (equal to $1 - 1/r$) to never return to the ball. Another explication follows from the theory of boundary value problems for elliptic differential operators: the exterior Neumann problem has a unique solution, while the solution of the interior Neumann problem is defined up to a constant. Consequently, the Dirichlet-to-Neumann $\mathcal{M}$ operator should be invertible for the exterior problem that implies a simple condition for its eigenvalues: $\mu_\alpha \neq 0$. On the contrary, $\mathcal{M}$ is not invertible for the interior problem providing the condition that at least one eigenvalue is zero.

4 Stochastic Approaches to Laplacian Transport Phenomena

In this section, we return to the Laplacian transport phenomena, discussed at the beginning. First, we are going to introduce the notion of source, diffusing particles started from. The definition of the partially reflected Brownian motion will require only a minor modification. After that, the recently developed continuous approach will be presented with a special emphasis on its physical importance. Finally, we will mention two other physical descriptions which can be now considered as useful approximations to the continuous approach.

4.1 Notion of Source

The description of the partially reflected Brownian motion given in the previous section does not involve a source, an important element for the Laplacian transport phenomena. In this subsection, we are going to discuss the extension of previous definitions in order to introduce the source. As one will see, a minor modification will be sufficient.

Throughout this subsection, we consider a bounded domain Ω with twice continuous differentiable boundary composed of two disjoint parts, $\partial\Omega$ and $\partial\Omega_0$, referred as *working interface* and *source* respectively[3]. In practice, the working interface and the source are well separated in space, therefore one may think about a circular ring as generic domain.

As previously, one considers the reflected Brownian motion $\hat{W}_t$, started from any point $x \in \Omega \cup \partial\Omega \cup \partial\Omega_0$ and reflected on the whole boundary $\partial\Omega \cup \partial\Omega_0$, the corresponding local time process ℓ_t, and the stopping time $\mathbb{T}_\Lambda^x$ defined in 3.1. Let us introduce a new stopping time τ as the first moment when the process $\hat{W}_t$ hits the source $\partial\Omega_0$:

$$\tau = \inf\{t > 0 \ : \ \hat{W}_t \in \partial\Omega_0\}$$

Then, the spread harmonic measure can be defined for any subset A from Borel σ-algebra $\mathcal{B}(\partial\Omega)$ (defined on the working interface!) as

$$\omega_{x,\Lambda}\{A\} = \mathbb{P}\{\hat{W}_{\mathbb{T}_\Lambda^x} \in A, \ \mathbb{T}_\Lambda^x < \tau < \infty\} \tag{22}$$

We remark two distinctions with respect to the previous definition 3.2:

- The measure is considered on Borel subsets of the working interface $\partial\Omega$ only, whereas the reflected Brownian motion $\hat{W}_t$ and the local time process ℓ_t are defined on the whole boundary $\partial\Omega \cup \partial\Omega_0$.
- There is a supplementary condition $\mathbb{T}_\Lambda^x < \tau$ providing that the process $\hat{W}_t$ should be absorbed (killed) on the working interface before hitting the source.

Note that $1 - \omega_{x,\Lambda}\{\partial\Omega\}$ is the probability that the process $\hat{W}_t$ started from a given point $x \in \Omega$ hits the source $\partial\Omega_0$ before its final absorption on the working interface $\partial\Omega$.

One can easily extend the lemma 3.3 to this spread harmonic measure:

Lemma 4.1 *For any subset A from $\mathcal{B}(\partial\Omega)$ and fixed positive Λ, the spread harmonic measure $\omega_{x,\Lambda}\{A\}$, considered as function of x, solves the boundary value problem:*

$$\Delta\omega_{x,\Lambda}\{A\} = 0 \quad (x \in \Omega), \qquad \begin{array}{ll} \left[I - \Lambda\dfrac{\partial}{\partial n}\right]\omega_{x,\Lambda}\{A\} = \mathbb{I}_A(x) & (x \in \partial\Omega) \\ \omega_{x,\Lambda}\{A\} = 0 & (x \in \partial\Omega_0) \end{array} \tag{23}$$

Proof is similar to those of the lemma 3.3. The last condition holds since $x \in \partial\Omega_0$ implies $\tau = 0$. □

Corollary 4.2 *Function $C_\Lambda(x) = C_0(1 - \omega_{x,\Lambda}\{\partial\Omega\})$ solves the boundary value problem (1ñ3).*

Proof is a direct verification. □

Consequently, a simple introduction of the source allows to apply the previous description of the partially reflected Brownian motion to study the Laplacian transport phenomena. Due to reversibility of the Brownian motion, one may think that $(1 - \omega_{x,\Lambda}\{\partial\Omega\})dx$ gives also the probability to find the partially reflected Brownian motion, started from the absorbing source, in dx vicinity of the point $x \in \Omega$, under partially absorbing condition on the

[3] Previously, the whole boundary of the domain had been considered as the working interface. For this reason, we preserve the same notation $\partial\Omega$ for this object and hope that it will not lead to ambiguities.

working interface $\partial\Omega$. Note that such way of reasoning, being intuitive and useful, is quite formal. In particular, the (simple) Brownian motion started from the source returns to it infinitely many times with probability 1. If one really needs to define such a process, the starting point should be taken slightly above the source.

Since the boundary $\partial\Omega$ is always supposed to be smooth, one can introduce the spread harmonic measure density $\omega_{x,\Lambda}(s)$. In turn, the kernel of the spreading operator is defined as previously, $T_\Lambda(s, s') \equiv \omega_{s,\Lambda}(s')$. In particular, one retrieves the relation (11):

$$\omega_{x,\Lambda}(s) = \int\limits_{\partial\Omega} ds' \omega_{x,0}(s')\, T_\Lambda(s', s)$$

where the harmonic measure density $\omega_{x,0}(s)$ is defined by relation (22) with $\Lambda = 0$.

The definition of the Dirichlet-to-Neumann operator can be also extended to domains with a source. For a given function $f \in H^1(\partial\Omega)$, one solves the Dirichlet problem in the domain Ω:

$$\Delta u = 0 \quad (x \in \Omega), \qquad u = f \quad (x \in \partial\Omega), \qquad u = 0 \quad (x \in \partial\Omega_0)$$

In principle, one could consider another function on the source. For a given function f on $\partial\Omega$, the Dirichlet-to-Neumann operator $\mathcal{M}$, acting from $H^1(\partial\Omega)$ to $L^2(\partial\Omega)$, associates the new function $g = \partial u/\partial n$ on the working interface $\partial\Omega$. One demonstrates general properties of this operator and its relation to the partially reflected Brownian motion in a straight way. In particular, the spreading operator T_Λ, defined by its kernel $T_\Lambda(s, s')$, coincides with the resolvent operator $[I + \Lambda\mathcal{M}]^{-1}$. However, some normalization properties may be changed. In particular, for the probability kernel $T_\Lambda(s, s')$, one has

$$\int\limits_{\partial\Omega} T_\Lambda(s, s') ds' < 1$$

since the PRBM started from the working interface $\partial\Omega$ can be absorbed on the source.

4.2 Continuous Approach

The stochastic treatment by means of the partially reflected Brownian motion provides the solution of the problem (1–3) describing the Laplacian transport phenomena: $C_\Lambda(x) = C_0(1 - \omega_{x,\Lambda}\{\partial\Omega\})$ (see corollary 4.2). One can go further using the close relation to the Dirichlet-to-Neumann operator [33]. According to the lemma 4.1, the density $\omega_{x,\Lambda}\{\partial\Omega\}$, considered as function of x, solves the boundary value problem:

$$\Delta\omega_{x,\Lambda}\{\partial\Omega\} = 0 \quad (x \in \Omega), \qquad \left[I - \Lambda\frac{\partial}{\partial n}\right]\omega_{x,\Lambda}\{\partial\Omega\} = 1 \quad (x \in \partial\Omega), \qquad \omega_{x,\Lambda}\{\partial\Omega\} = 0 \quad (x \in \partial$$

The restriction of the function $\omega_{x,\Lambda}\{\partial\Omega\}$ on $\partial\Omega$ can be written with the help of the Dirichlet-to-Neumann operator $\mathcal{M}$ as

$$\omega_{x,\Lambda}\{\partial\Omega\} = \left[(I + \Lambda\mathcal{M})^{-1}1\right](s) = [T_\Lambda 1](s)$$

where $1(s)$ stands for a constant (unit) function on the working interface.

One defines then the flux density $\phi_\Lambda(s)$ across the working interface $\partial\Omega$:

$$\phi_\Lambda(s) = D\,\frac{\partial C_\Lambda}{\partial n}(s) = -D\,\frac{\partial \omega_{x,\Lambda}\{\partial\Omega\}}{\partial n}(s)$$

Since the normal derivative of a harmonic function can be represented as the application of the Dirichlet-to-Neumann operator to the restriction of this function on the boundary, one writes

$$\phi_\Lambda(s) = D\,\mathcal{M}\omega_{s,\Lambda}\{\partial\Omega\} = D\,\mathcal{M}[T_\Lambda 1](s)$$

(the sign is changed due to particular orientation of the normal derivative). Taking $\Lambda = 0$, one finds $\phi_0(s) = D\,[\mathcal{M}1](s)$ and finally

$$\phi_\Lambda(s) = [T_\Lambda \phi_0](s)$$

The transport properties of the working interface can be characterized by a physical quantity called *spectroscopic impedance*. We remind that the impedance of an electric scheme is defined as the tension applied between two external poles, divided by the total electric current passing through. The formal analogy between the electric problem and the diffusive transport, discussed in section 1, leads to a natural definition of the impedance in our case as the concentration C_0 on the source $\partial\Omega_0$ divided by the total flux across the working interface $\partial\Omega$:

$$Z_{cell}(\Lambda) = \frac{C_0}{\int\limits_{\partial\Omega} ds\;\phi_\Lambda(s)}$$

Taking $\Lambda = 0$, one deals with a purely absorbing interface $\partial\Omega$: any particle arrived to $\partial\Omega$ is immediately absorbed (without reflections). In other words, such interface has no resistance for passage across it. Consequently, the impedance $Z_{cell}(0)$ represents the "access resistance" by the volume: the possibility that the Brownian motion can return to the source without hitting the working interface. The resistance property of the working interface can be thus characterized by the difference between $Z_{cell}(\Lambda)$ and $Z_{cell}(0)$, called *spectroscopic impedance*:

$$Z_{sp}(\Lambda) = Z_{cell}(\Lambda) - Z_{cell}(0)$$

Using the simple identity

$$\left((\phi_\Lambda - \phi_0)\cdot 1\right)_{L^2} = \frac{\Lambda}{D}\left(\phi_\Lambda \cdot \phi_0\right)_{L^2} \qquad (24)$$

one writes the spectroscopic impedance as

$$Z_{sp}(\Lambda) = \frac{\Lambda}{D}\,\frac{\left(\phi_\Lambda \cdot \phi_0\right)_{L^2}}{\left(\phi_\Lambda \cdot 1\right)_{L^2}\left(\phi_0 \cdot 1\right)_{L^2}}$$

Using again the identity (24), one finds a more convenient form:

$$Z_{sp}(\Lambda) = \frac{1}{\dfrac{1}{Z(\Lambda)} - \dfrac{1}{Z_{cell}(0)}}$$

The new function

$$Z(\Lambda) = \frac{\Lambda}{D}\left(T_\Lambda \phi_0^h \cdot \phi_0^h\right)_{L^2} \qquad \phi_0^h(s) = \frac{\phi_0(s)}{\left(\phi_0 \cdot 1\right)_{L^2}}$$

can be called *effective impedance*, where $\phi_0^h(s)$ is the normalized flux density towards the perfectly absorbing working interface $\partial\Omega$. Finally, the spectral decomposition of the spreading operator T_Λ on the basis of eigenfunctions $\mathbf{V}_\alpha$ of the Dirichlet-to-Neumann operator $\mathcal{M}$ leads to the important relation for the effective impedance:

$$Z(\Lambda) = \frac{\Lambda}{D}\sum_\alpha \frac{F_\alpha}{1+\Lambda\mu_\alpha} \qquad F_\alpha = \left(\phi_0^h \cdot \mathbf{V}_\alpha\right)_{L^2}\left(\phi_0^h \cdot \mathbf{V}_\alpha^*\right)_{L^2} \tag{25}$$

This relation presents the central result of the continuous approach developed in [33]. Let us briefly discuss its physical meaning. The spectroscopic impedance $Z_{sp}(\Lambda)$ or, equivalently, the effective impedance $Z(\Lambda)$, is a physical quantity which characterizes the transport properties of the whole working interface. More importantly, this quantity can be measured directly in experiment (e.g., in electrochemistry). On the other hand, the local transport properties of the working interface are described by the single physical parameter Λ, being related to the membrane's permeability W, the electrode's resistance r or the catalyst's reactivity K (see Section 1). Varying the parameter Λ, one changes the local transport properties at each boundary point and, consequently, the whole linear response of the working interface. In a first sight, one may think that an increase of the local boundary resistance should imply a proportional increase of the whole boundary resistance, i.e., $Z(\Lambda) \sim \Lambda$. This reasoning being true for a planar surface becomes invalid in a general case due to geometrical irregularities. In fact, an irregular geometry modifies considerably the linear response of the working interface [22–28]. The boundary value problem (1–3) describing the Laplacian transport phenomena on the average allows, in principle, to study such geometrical influence. Practically, however, this is a very difficult problem. The continuous approach provides an efficient tool to carry out these studies both in theoretical and numerical ways. In particular, the relation (25) makes explicit the impedance dependence on the local transport properties (parameter Λ) and allows to identify contributions due to the *physics* and due to the *geometry*, involved in the problem in a complex manner. In other words, whatever the physical problem (diffusion across semi-permeable membranes, heterogeneous catalysis or electric transport), the geometry enters *only* through the spectral characteristics of the Dirichlet-to-Neumann operator $\mathcal{M}$: its eigenvalues μ_α and the spectral components F_α of the normalized flux density $\phi_0^h(s)$ on the basis of its eigenfunctions $\mathbf{V}_\alpha(s)$.

The other important meaning of the relation (25) can be outlined if one considers the inverse problem [33]: what is the most available information that one can retrieve from a measurement of the spectroscopic impedance of an unknown working interface? The mathematical response can be given immediately if one rewrites (25) as Laplace transform of the new function $\zeta(\lambda)$:

$$Z(\Lambda) = \frac{1}{D}\int_0^\infty d\lambda\, e^{-\lambda/\Lambda}\,\zeta(\lambda), \qquad \zeta(\lambda) \equiv \sum_\alpha F_\alpha e^{-\lambda\mu_\alpha}$$

Under assumption to be able to measure the impedance with an absolute precision, one can reconstruct the function $\zeta(\lambda)$ and, consequently, the set of characteristics $\{\mu_\alpha, F_\alpha\}$ which may be thus called *geometrical spectrum* of the working interface. In practice, however, the inverse problem is more difficult [62].

In two following subsections, we are going to discuss some aspects of the semi-continuous and the discrete descriptions of the Laplacian transport phenomena which can be referred as physical approaches based on the intuitive notion of partial reflections on the boundary. Since these descriptions turn out to be approximations to the continuous approach, we do not present the circumstantial details.

4.3 Semi-continuous Approach

The semi-continuous approach, developed by Halsey and Leibig [25], is based on the following construction (for details, see [33]). For a given domain Ω with smooth bounded boundary $\partial\Omega$, one considers the Brownian motion started from a point $x \in \Omega$. When it hits the boundary at some point s, two complementary events may happen:

- with probability ε, the Brownian motion is reflected to the interior point $s + an(s)$, slightly above the boundary (here $n(s)$ is the unit normal vector to the boundary at point s, a is a small positive parameter); the Brownian motion continues from this point;
- or, with probability $1-\varepsilon$, the Brownian motion is terminated at this point s (absorbed on the boundary).

This motion continues until the absorption on the boundary and can be called *Brownian motion re ected with jump* . Two new parameters, the jump distance a and the reflection probability ε, are related to the given physical parameter Λ [50]:

$$\varepsilon = \frac{1}{1 + (a/\Lambda)} \tag{26}$$

(one fixes Λ and takes $0 < a < \Lambda$ in order to have $0 < \varepsilon < 1$).

Now, one can calculate the probability $\omega_{x,\Lambda}^{(a)}(s)ds$ that this process is finally absorbed in ds vicinity of the boundary point s. Since the motions before and after each reflection are independent, this probability can be obtained as the sum of probabilities to be absorbed after 0, 1, 2, ... reflections:

$$\omega_{x,\Lambda}^{(a)}(s)ds = \big[\omega_x(s)ds\big]\,(1-\varepsilon) + \int\limits_{\partial\Omega} \big[\omega_x(s_1)ds_1\big]\,\varepsilon\,\big[\omega_{s_1+an(s_1)}(s)ds\big]\,(1-\varepsilon) +$$

$$+ \int\limits_{\partial\Omega}\int\limits_{\partial\Omega} \big[\omega_x(s_1)ds_1\big]\,\varepsilon\,\big[\omega_{s_1+an(s_1)}(s_2)ds_2\big]\,\varepsilon\,\big[\omega_{s_2+an(s_2)}(s)ds\big]\,(1-\varepsilon) + \,\ldots$$

For example, the third term represents the probability to hit the boundary in ds_1 vicinity of the point s_1, to be reflected to the neighboring point $s_1+an(s_1)$, to hit again the boundary in ds_2 vicinity of the point s_2, to be reflected to the neighboring point $s_2 + an(s_2)$, to hit the boundary for the last time in ds vicinity of the point s, and to be finally absorbed. Introducing the integral operator $Q^{(a)}$, acting from $L^2(\partial\Omega)$ to $L^2(\partial\Omega)$ as

$$[Q^{(a)}f](s) = \int\limits_{\partial\Omega} ds'\, f(s')\, \omega_{s'+an(s')}(s)$$

one rewrites the previous sum as the application of the new integral operator $T_\Lambda^{(a)}$ to the harmonic measure density $\omega_x(s)$:

$$\omega_{x,\Lambda}^{(a)}(s) = [T_\Lambda^{(a)}\omega_x](s) \qquad\qquad T_\Lambda^{(a)} = (1-\varepsilon)\sum_{k=0}^{\infty}\left(\varepsilon Q^{(a)}\right)^k \tag{27}$$

What happens when the jump distance a goes to 0? Hitting the boundary, the Brownian motion will be reflected to interior points lying closer and closer to the boundary, i.e., displacements of the Brownian motion between two serial hits are getting smaller and smaller. At the same time, the reflection probability ε tends to 1 according to relation (26), i.e., the average number of reflections increases. Indeed, the distribution of the random number $\mathcal{N}$ of reflections until the final absorption is simply

$$\mathbb{P}\{\mathcal{N} = n\} = (1-\varepsilon)\,\varepsilon^n \tag{28}$$

implying that the average number $<\mathcal{N}>= \varepsilon(1-\varepsilon)^{-1}$ goes to infinity. Does the limiting process exist? The situation is complicated by the *local* choice between reflection and absorption: at each hitting point, the motion can be absorbed with vanishing probability $1-\varepsilon$. In order to get round this difficulty, one can consider the process from a slightly different point of view. Actually, one can replace the local condition of the absorption (with probability $1-\varepsilon$) by its *global* analog: the process is absorbed on the boundary when the number of reflections exceeds a random variable $\mathcal{N}$ distributed according to the geometrical law (28). Evidently, this modification does not change at all the properties of the process. At the same time, we gain that the condition of the absorption is independent of the Brownian motion between serial hits. As a consequence, we can consider the corresponding limits (as $a \to 0$) separately. So, the Brownian motion reflected with jump should tend to the reflected Brownian motion as the jump distance a vanishes. This motion, however, is conditioned to stop when the number of reflections exceeds the random variable $\mathcal{N}$. Since the average number $<\mathcal{N}>$ goes to infinity in the limit $a \to 0$, it is convenient to consider a normalized variable $\chi = a\mathcal{N}$ obeying the following distribution:

$$\mathbb{P}\{\chi \geq \lambda\} = \mathbb{P}\{\mathcal{N} \geq \lambda/a\} = \sum_{[\lambda/a]}^{\infty} \mathbb{P}\{\mathcal{N} = n\} \simeq \varepsilon^{[\lambda/a]} \simeq \exp[-\lambda/\Lambda] \tag{29}$$

(the last equality is written with the help of (26) when a goes to 0). Since the number of reflections on jump distance a, multiplied by a, tends to the local time process according to Lévy's formula (9), the previous condition of absorption can be reformulated: the motion is absorbed when its local time process exceeds a random variable distributed according to the exponential law (29). One thus concludes that the Brownian motion reflected with jump should tend to the partially reflected Brownian motion defined in 3.1.

The above analysis, presented as a sketch (without proofs), does not pretend to a mathematical rigour. It may be considered rather as a possible justification which can be brought for the semi-continuous approach if necessary. In particular, one can demonstrate that the density $\omega_{x,\Lambda}^{(a)}(s)$, given by relation (27), tends to the spread harmonic measure density as the jump distance a goes to 0:

$$\omega_{x,\Lambda}(s) = \lim_{a\to 0} \omega_{x,\Lambda}^{(a)}(s)$$

This relation may be useful for numerical computations. Similarly, the integral operator $T_\Lambda^{(a)}$ should converge to the spreading operator T_Λ as $a \to 0$. Calculating the geometrical series in (27) and representing $T_\Lambda^{(a)}$ as

$$T_\Lambda^{(a)} = (1-\varepsilon)\left(I - \varepsilon Q^{(a)}\right)^{-1} = \left(I + \Lambda\, \frac{I - Q^{(a)}}{a}\right)^{-1}$$

one obtains the following approximation for the Dirichlet-to-Neumann operator:

$$\mathcal{M} = \lim_{a\to 0} \frac{I - Q^{(a)}}{a}$$

Again, this relation may be useful for the numerical computation of the left hand side.

The advantages of the semi-continuous approach are based on an apparent intuitive meaning of partial reflections on the boundary. Moreover, this approach provides even a more realistic description of physico-chemical processes on microscopic level. For example, if one considers the partially reflected Brownian motion started from a boundary point, the number of hits of the boundary is infinite for any moment $t > 0$ that sounds impossible for real physical species. The keypoint is that, for diffusion across a semi-permeable membrane or heterogeneous reaction on a catalytic surface, the description by the boundary value problem (1–3) cannot be justified on scales less than the mean free path of diffusing particles. Since the continuous limit $a \to 0$ requires such *non-physical* scales, it is not surprising that the limiting process (the PRBM) presents some irrealistic properties from the physical point of view. The similar limitation happens for the electric transport problem when the smallest physical scale is given by the thickness of the double layer, being close to the Debye-Hückel length [24, 25]. Evidently, this remark does not devaluate the efficiency of the continuous approach based on the partially reflected Brownian motion. On the contrary, the mathematical rigour of this approach justifies the semi-continuous description and simplifies its study by introducing the Dirichlet-to-Neumann operator. However, dealing with a mathematical description of a physical problem, one should take care that deduced consequences do not go beyond the ranges of the model.

The capabilities of the semi-continuous approach are essentially limited by the fact that the central operator $Q^{(a)}$ is not self-adjoint (the function $\omega_{s+an(s)}(s')$ is not symmetric with respect to the permutation of s and s' except specific cases). As a consequence, one cannot develop the spectral decomposition (25) of the impedance, i.e., there is no possibility to distinguish contributions of different modes. Although the operator $Q^{(a)}$ is defined naturally by the harmonic measure density, it does not provide a proper description of the problem as it was done with the Dirichlet-to-Neumann operator.

4.4 Discrete Approach

Another stochastic approach to Laplacian transport phenomena has been developed by Filoche and Sapoval [20]. The main idea is to model the partially reflected Brownian motion by random walks on a lattice with partial reflections on the boundary. Actually, one discretizes a given domain Ω by d-dimensional hypercubic lattice of mesh a and considers the following stochastic process: started from a remote source, a random walker jumps to a neighboring site at each step with probability $(2d)^{-1}$. When the walker arrives to a boundary site, it can be reflected to its neighboring site (belonging to the bulk) with probability ε (and the motion continues), or it can be absorbed with probability $(1-\varepsilon)$. The motion continues until the final absorption on the boundary, or the return to the source. One can show [50] that the discrete parameters a and ε are related by the expression (26) involving the continuous physical parameter Λ.

In the discrete description, the harmonic measure density is replaced by the distribution of hitting probabilities $(\mathbf{P}_0)_j$ of boundary sites j, (simple) random walks being started from a remote source. Let $Q_{j,k}^{(a)}$ denote the probability to arrive to the boundary site k starting from the boundary site j by a random walk in the bulk without hitting the boundary or the source during the walk[4]. One can thus calculate the distribution of probabilities $(\mathbf{P}_\Lambda)_j$ to be finally absorbed on the boundary site j, when random walks with partial reflections are started from a remote source. Indeed, the Markov property of this process allows to calculate $(\mathbf{P}_\Lambda)_j$ as the sum of contributions provided by random trajectories with $0, 1, 2, ...$ reflections before the final absorption:

$$(\mathbf{P}_\Lambda)_j = (\mathbf{P}_0)_j(1-\varepsilon)+\sum_{k_1}(\mathbf{P}_0)_{k_1}\varepsilon Q_{k_1,j}^{(a)}(1-\varepsilon)+\sum_{k_1}\sum_{k_2}(\mathbf{P}_0)_{k_1}\varepsilon Q_{k_1,k_2}^{(a)}\varepsilon Q_{k_2,j}^{(a)}(1-\varepsilon)+...$$

(we remind that $\varepsilon = (1+a/\Lambda)^{-1}$). For example, the second term represents the product of the following probabilities: to hit a boundary site k_1, to be reflected to its neighboring site, to arrive to the boundary site j, and to be finally absorbed on it. If one considers $\mathbf{P}_0$ and $\mathbf{P}_\Lambda$ as vectors and $Q^{(a)}$ as matrix, the summation over intermediate sites $k_1, k_2, ...$ can be understood as matrix product:

$$\mathbf{P}_\Lambda = \left[(1-\varepsilon)\sum_{n=0}^{\infty}\left(\varepsilon Q^{(a)}\right)^n\right]\mathbf{P}_0$$

i.e., the distribution of absorption probabilities $(\mathbf{P}_\Lambda)_j$ is obtained as the application of a linear operator, depending on $Q^{(a)}$ and Λ (or ε), to the distribution of hitting probabilities $(\mathbf{P}_0)_j$. The symmetric matrix $Q^{(a)}$ represents a self-adjoint operator, called *Brownian self-transport operator*. Using the normalization property $|Q^{(a)}| \le 1$ and relation (26) between Λ and ε, one obtains:

$$\mathbf{P}_\Lambda = T_\Lambda^{(a)}\mathbf{P}_0 \qquad T_\Lambda^{(a)} = \left[I + \Lambda\,\frac{I-Q^{(a)}}{a}\right]^{-1}$$

[4]We use the same notation $Q^{(a)}$ for the integral operator in semi-continuous approach and for the matrix of these probabilities since they have the same meaning and even may be used to approximate each other.

The operator $T_\Lambda^{(a)}$, depending on the lattice parameter a, is called *(discrete) spreading operator*. The previous relation, written explicitly as

$$(\mathbf{P}_\Lambda)_j = \sum_k (\mathbf{P}_0)_k \big(T_\Lambda^{(a)}\big)_{k,j}$$

allows to separate random trajectories in two independent parts:

- the random walker started from a remote source arrives to the boundary site k (first factor);
- it continues the motion with partial reflections until the final absorption on the boundary site j (second factor).

One concludes that the absorption probabilities $(\mathbf{P}_\Lambda)_j$ provide a discrete analog of the spread harmonic measure density, while the matrix $\big(T_\Lambda^{(a)}\big)_{k,j}$ is a discrete analog of the kernel $T_\Lambda(s, s')$ of the spreading operator $T_\Lambda = [I + \Lambda\mathcal{M}]^{-1}$. In particular, the bounded operators $(I - Q^{(a)})/a$ can be understood as discrete approximations of the Dirichlet-to-Neumann operator $\mathcal{M}$ (in resolvent sense). As for the semi-continuous approach, we do not furnish the corresponding proofs (see [33] for more details).

The advantage of the discrete description with respect to the semi-continuous approach is based on the fact that the Brownian self-transport operator $Q^{(a)}$ and, consequently, the (discrete) spreading operator $T_\Lambda^{(a)}$ are self-adjoint. This property allows to employ all the machinery of the spectral theory in order to express the physical characteristics of the Laplacian transport through eigenmodes of this operator in an explicit way. For example, the spectral decomposition (25) can be also written in the discrete case. Such decompositions have been used to study the Laplacian transport towards irregular geometries [33, 50]. Moreover, the discrete description provides at least two different ways to study the problem numerically: direct Monte Carlo simulations and discrete boundary elements method [63].

The discrete description, being intuitively the most simple and useful, may lead to mathematical difficulties when one tries to proceed the continuous limit a going to 0. Although the partially reflected Brownian motion is the natural limit of random walks with partial reflections, its rigorous demonstration, in our knowledge, is not yet realized in details. The interested reader can find more information on this topic in [64–69].

5 Conclusion

The application of stochastic processes to represent the solution of boundary value problems is well known and wide used. In particular, Monte Carlo simulations are generally based on this technique. In this paper, we gave an overview of the Laplacian transport phenomena found in different scientific domains (e.g., physics, electrochemistry, chemistry, physiology) and related stochastic approaches to describe them. The most attention has been paid to the recently developed continuous approach based on the partially reflected Brownian motion. This stochastic process can be thought as rigorous mathematical description for random trajectories of diffusing particles hitting a semi-permeable interface, in comparison with more intuitive physical descriptions by semi-continuous and discrete approaches. The partially reflected Brownian motion turns out to be the natural limit of the Brownian motion reflected with jump (semi-continuous approach) and of the random walks with partial reflections (discrete approach).

The profound relation between the partially reflected Brownian motion and spectral properties of the Dirichlet-to-Neumann operator $\mathcal{M}$ are found to be useful for practical purposes. In particular, the kernel of the resolvent operator $T_\Lambda = [I + \Lambda\mathcal{M}]^{-1}$ provides the probability density $T_\Lambda(s, s')$ allowing to reconstruct the spread harmonic measure density $\omega_{x,\Lambda}(s)$. Moreover, the spectral decomposition on the complete basis of eigenfunctions of the Dirichlet-to-Neumann operator leads to the explicit analytical formula for this density. Consequently, the use of the operator $\mathcal{M}$ gives an efficient way to study different probability distributions related to the partially reflected Brownian motion.

The spectral decomposition of the spectroscopic impedance, characterizing the linear response of the whole working interface, provides an explicit analytical dependence on the physical parameter Λ allowing to identify *physical* and *geometrical* contributions which were involved in a complex manner. The geometrical spectrum of the working interface contains the complete information about its transport properties. The combined use of stochastic characteristics of the partially reflected Brownian motion and spectral properties of the Dirichlet-to-Neumann operator opens encouraging possibilities for further understanding of various physical and chemical transport processes in nature. In this light, a more profound mathematical analysis of these objects seems to be an important perspective for the present study.

Acknowledgement

The author thanks Professor B. Sapoval and M. Filoche for valuable discussions and fruitful collective work on physical aspects of the Laplacian transport phenomena.

References

[1] E.R. Weibel, The Pathway for oxygen. Structure and function in the mammalian respiratory system, Harvard University Press, Cambridge, Massachusetts and London, England, 1984.

[2] B. Mauroy, M. Filoche, E.R. Weibel, B. Sapoval, *Nature* **427**, 633-636, 2004.

[3] B. Mauroy, Hydrodymanique dans les poumons, influence entre flux et géométries (Hydrodymanics in the Lung, Interplay between Flow and Geometries), PhD thesis at ENS de Cachan, 2004.

[4] B. Sapoval, M. Filoche and E. Weibel, in "*Branching in nature*" Ed. by V. Fleury, J.-F. Gouyet and M. Leonetti, EDP Sciences/Springer Verlag, 225-242, 2001.

[5] B. Sapoval, M. Filoche, E.R. Weibel, *Proc. Nat. Ac. Sc.* **99** (16), 10411-10416, 2002.

[6] B. Sapoval, E.R. Weibel, M. Filoche, in "*Fractals in Biology and Medecine*", Eds. G.A. Losa, D. Merlini, T.F. Nonennmacher, E.R. Weibel, 25-38, 2002.

[7] M. Felici, M. Filoche, B. Sapoval, *J. Appl. Physiol.* **94**, 2010-2016, 2003.

[8] M. Felici, B. Sapoval, M. Filoche, *Phys. Rev. Lett.* **92**, 068101, 2004.

[9] M. Felici, *Physique du transport diffusif de l'oxygene dans le poumon humain* (Physics of the Diffusive Transport of the Oxygen in the Human Lung), PhD Thesis at Ecole Polytechnique, France, 2003.

[10] D.S. Grebenkov, M. Felici, M. Filoche, B. Sapoval (submitted to *Phys. Rev. Lett.*)

[11] B. Haefeli-Bleurer, E.R. Weibel, *Anat. Rec.* **220**, 401, 1988.

[12] B. Sapoval, J.S. Andrade Jr., M. Filoche, *Chem. Eng. Science* **56**, 5011, 2001.

[13] J.S. Andrade jr., M. Filoche, B. Sapoval, *Europhys. Lett.* **55** (4), 573-579, 2001.

[14] J.S. Andrade jr., H.F. da Silva, M. Baquil, B. Sapoval, *Phys. Rev. E* **68**, 041608, 2003.

[15] M. Filoche, J.S. Andrade Jr., B. Sapoval, Proceedings of LAWNP'03, Salvadore da Bahia, 28 Sept. – 3 Oct. 2003, to be published in *Physica A*.

[16] M. Filoche, J.S. Andrade Jr., B. Sapoval (submitted to *AICHE*).

[17] K. Malek, M.-O. Coppens, *Phys. Rev. Lett.* **87**, 125505, 2001.

[18] B. Sapoval, *Phys. Rev. Lett.* **73**, 3314, 1994.

[19] B. Sapoval, in "*Fractals and Disordered Systems*", Eds. A. Bunde, S. Havlin, Springer, 233-261, 1996.

[20] M. Filoche, B. Sapoval, *Eur. Phys. J. B* **9**, 755-763, 1999.

[21] I. Wolff, *Phys. Rev.* **27**, 755-763, 1926.

[22] R. de Levie, *Electrochimica Acta* **10**, 113-130, 1965.

[23] L. Nyikos, T. Pajkossy, *Electrochimica Acta* **30**, 1533-1540, 1985.

[24] T.C. Halsey, *Phys. Rev. A* **35**, 3512, 1987.

[25] T.C. Halsey, M. Leibig, *Annals of Physics* **219**, 109, 1992.

[26] B.Sapoval, *Solid State Ionics* **23**, 253-259, 1987.

[27] B. Sapoval, J.-N. Chazalviel, J. Peyriere, *Phys. Rev. A* **38**, 5867, 1988.

[28] B. Sapoval, E. Chassaing, *Physica A* **157**, 610-617, 1989.

[29] M. Filoche, B. Sapoval, *J. Phys. I France* **7**, 1487-1498, 1997.

[30] B. Sapoval, M. Filoche, K. Karamanos, R. Brizzi, *Eur. Phys. J. B* **9**, 739-753, 1999.

[31] M. Filoche, B. Sapoval, *Phys. Rev. Lett.* **84** (25), 5776-5779, 2000.

[32] M. Filoche, B. Sapoval, *Electrochimica Acta* **46**, 213-220, 2000.

[33] D.S. Grebenkov, Transport laplacien aux interfaces irrégulieres : Etude théorique, numérique et expérimentale (Laplacian Transport towards Irregular Interfaces: A Theoretical, Numerical and Experimental Study), PhD thesis at Ecole Polytechnique.

[34] M. Freidlin, Functional Integration and Partial Differential Equations, Annals of Mathematics Studies, Princeton University Press, Princeton, New Jersey, 1985.

[35] A.N. Borodin, P. Salminen, *Handbook of Brownian Motion: Facts and Formulae*, Birkhauser Verlag, Basel-Boston-Berlin, 1996.

[36] K. Ito, H.P. McKean, *Diffusion Processes and Their Sample Paths*, Springer-Verlag, Berlin, 1965.

[37] P. Lévy, *Processus Stochastiques et Mouvement Brownien*, Paris, Gauthier-Villard, 1948-1965.

[38] R.F. Bass, *Diffusions and Elliptic Operators*, Springer, 1998.

[39] S.C. Port, C.J. Stone, *Brownian Motion and Classical Potential Theory*, New York, Academic Press, 1978.

[40] D.R.J. Revuz, M. Yor, *Continuous Martingales and Brownian Motion*, Berlin, Springer, 1991, 1999.

[41] J.B. Garnett, D.E. Marshall, *Harmonic Measure*, Cambridge University Press (to appear).

[42] J.B. Garnett, *Applications of Harmonic Measure*, John Wiley & Sons, New York, 1986.

[43] N.G. Makarov, *St. Petersburg Math. J.* **10** (2), 217-267, 1999.

[44] G.F. Roach, *Green's functions*, 2nd Ed., Cambridge University Press, 1982.

[45] R.P. Feynman, A.R. Hibbs, *Quantum Mechanics and Path Integrals*, New York : McGraw-Hill, 1965.

[46] R.F. Anderson, S. Orey, *Nagoya Math. J.* **60**, 189-216, 1976.

[47] P.L. Lions, A.S. Sznitman, *Comm. Pure Appl. Math.* **37**, 511-537, 1984.

[48] Y. Saisho, *Probab. Theory Rel. Fields* **74**, 455-477, 1987.

[49] S. Kakutani, *Imp. Acad. Tokyo* **20**, 706-714, 1944.

[50] D.S. Grebenkov, M. Filoche, B. Sapoval, *Eur. Phys. J. B* **36** (2), 221-231, 2003.

[51] M.S. Agranovich, in *ìPartial Differential Equations IXî*, Ed. by M.S. Agranovich, Yu.V. Egorov, M.S. Shubin, *EMS* **79**, Springer, 1997.

[52] M.S. Birman, M.Z. Solomyak, *Spectral Theory of Self-Adjoint Operators in Hilbert Space*, D.Reidel Publishing Company, 1987.

[53] L. Hörmander, *The Analysis of Linear Partial Differential Operators* , 4 vol.; Berlin : Springer, 1983-1985.

[54] Yu. Egorov, *Pseudo-differential Operators, Singularities, Applications* , Basel, Boston, Berlin : Birkhauser, 1997.

[55] N. Jacob, *Pseudo-differential Operators and Markov Processes* , Berlin : Akademie-Verlag, 1996.

[56] B.-W. Schulze, *Pseudo-differential Operators on Manifolds with Singularities* , Amsterdam : North-Holland, 1991.

[57] M.E. Taylor, *Pseudodifferential Operators* , Princeton, New Jersey : Princeton University Press, 1981.

[58] M.E. Taylor, *Pseudodifferential Operators and Nonlinear PDE* , Boston : Birkhauser, 1991.

[59] P. Grisvard, *Elliptic Problems in Nonsmooth Domains* , *MSM* **24**, Pitman Advanced Publishing Program, 1985.

[60] P. Grisvard, *Singularities in Boundary Value Problems* , *RMA* **22**, Masson, Springer-Verlag, 1992.

[61] P.A.M. Dirac, *The Principles of Quantum Mechanics*, Oxford : Clarendon Press, 1958.

[62] D.S. Grebenkov, M. Filoche, B. Sapoval, (to be published)

[63] D.S. Grebenkov, in "*Focus on Probability Theory*", Volume 2, Nova Science Publishers Inc. (to be published)

[64] K.K. Sabelfeld, *Monte Carlo Methods in Boundary Value Problems*, Springer-Verlag: New York - Heidelberg, Berlin, 1991.

[65] K.K. Sabelfeld, N.A. Simonov, *Random Walks on Boundary for Solving PDEs* , Utrecht, The Netherlands, 1994.

[66] G.N. Milshtein, *Numerical Integration of Stochastic Differential Equations* , Kluwer, Dordrecht, the Netherlands, 1995.

[67] G.N. Milshtein, *Math. Comput. Simulation* **38**, 77-85, 1995.

[68] D.W.Strook, S.R.S. Varadhan, *Comm. Pure Appl. Math.* **24**, 147-225, 1971.

[69] C. Costantini, B. Pacchiarotti, F. Sartoretto, *SIAM J. Appl. Math.* **58** (1), 73-102, 1998.

In: Focus on Probability Theory
Editor: Louis R. Velle, pp. 171-190

ISBN 1-59454-474-3

Chapter 9

Occupation Time on Hypersurfaces for Diffusion Jump Processes and Variations of Characteristic Functions

Nikolai Dokuchaev
Department of Mathematics and Statistics,
University of Limerick, Ireland

Abstract

The paper studies occupation time on hypersurface and its variations for n-dimensional jump diffusion processes. It is shown that the occupation time on hypersurfaces does exists for the jump processes as a limit in variance for a wide class of piecewise smooth hypersurfaces, including some fractal type and moving surfaces. Solvability and uniqueness of integro-differential Kolmogorov–Fokker–Planck with generalized functions in coefficients are investigated. An analog of the Meyer-Tanaka formula is presented. For a process with varying drifts, diffusions, and jump parameters, formulas for variations of expectations of functionals of occupation time are derived, including variations of the characteristic functions.

MSC: 60J55, 60J60

Keywords: diffusion jump processes, occupation time on hypersurfaces, Kolmogorov–Fokker–Planck equations, Meyer-Tanaka formula, variations of functionals

1 Introduction

The occupation time or the occupation time of stochastic processes have been studied by many authors (see e.g. bibliography in survey papers of Geman and Horovitz (1980), Borodin (1989), Protter (1990)). For example, the occupation time has been studied for the scalar Brownian motion and scalar semimartingales (see e.g. Meyer (1976)) and survey in Protter (1990)), for general one-dimensional diffusions (Salminen (1993)), for stable processes (Nolan (1979)).

For vector continuous processes, the distribution of the occupation time is also well studied. McGill (1982) has derived an analog of Tanaka formula for a solution of a scalar homogeneous nonlinear diffusion equation. Bass (1983) investigated occupation time of multidimensional non-Markovian continuous semimartingales of general type and proved the existence of occupation time. Rosen and Yor (1991) had considered the occupation time for processes in the plain at points of intersections. Dokuchaev (1999) studied occupation time for degenerating diffusion vector processes.

It appears that many important properties of Brownian occupation time do not hold for a case the occupation time of n-dimensional diffusion processes on $(n-1)$-dimensional hypersurfaces. For example, this occupation time can not be presented as occupational measure in a case $n>1$. Hence it is not a perfect analog of Brownian occupation time.

In recent years, a number of variational and control problems involving local time has received a considerable attention (see, e.g., Benes and Shepp (1980), Harrison and Taylor (1978), Karatzas and Shreve (1984), Davis and Norman (1990)).

The paper studies occupation time on hypersurface for Markov n-dimensional jump processes and variational problems which involves occupation time. It could be useful for further treating of practically important problems such as in the cited papers of Benes and Shepp (1980), Harrison and Taylor (1978), Karatzas and Shreve (1984), Davis and Norman (1990). As is known, analogs of Kolmogorov–Fokker–Planck equations for the probability distribution for jump processes are second order integro-differential equations (see e.g. Carrany and Menaldy (1992)). We extend solvability and uniqueness results for these equations for a case when there are generalized functions in coefficients (Sections 3 and 4). A class of acceptable surfaces is described by Lemma 5.1; this class includes piecewise C^1–smooth $(n-1)$–dimensional surfaces at $\mathbf{R}^n$. Then we show that the occupation time exists as a limit in variance, and an analog of the Meyer-Tanaka formula is derived, i.e. the occupation time is presented as a stochastic integral (Section 6). Equations for the characteristic function of the occupation time are derived in Section 7. In Section 8, we present explicit formulas for variations of functionals of occupation time on surface for general diffusion process with varying drift and diffusion (Theorem 8.1).

2 Definitions

Consider a standard probability space $(\Omega,\mathcal{F},\mathbf{P})$, a standard n-dimensional Wiener process with independent components such that $w(0)=0$, and a Poisson measure $\nu(\cdot,t)$ in $\mathbf{R}^m$ such that the process $w(\cdot)$ and the measure $\nu(\cdot,t)$ are independent of each other. We assume that there exists a measure $\Pi(\cdot)$ in $\mathbf{R}^m$ such that $\mathbf{E}\nu(A,t)=t\Pi(A)$ for each measurable set $A\subset\mathbf{R}^m$.

Set $\widetilde{\nu}(A,t)\stackrel{\Delta}{=}\nu(A,t)-t\Pi(A)$.

Let a be a random real n-dimensional vector such that a does not depend on $w(\cdot)$ and $\nu(\cdot)$. We assume also that $\mathbf{E}|a|^2<+\infty$.

We consider an n-dimensional stochastic differential equation

$$\begin{cases} dy(t)=f(y(t),t)dt+\beta(y(t),t)dw(t)+\int_{\mathbf{R}^m}\theta(y(t),u,t)\widetilde{\nu}(du,dt), \\ y(0)=a. \end{cases} \tag{1}$$

The functions $f(x,t):\mathbf{R}^n\times\mathbf{R}\to\mathbf{R}^n$, $\beta(x,t):\mathbf{R}^n\times\mathbf{R}\to\mathbf{R}^{n\times n}$ and $\theta(x,u,t):\mathbf{R}^n\times\mathbf{R}^m\times\mathbf{R}\to\mathbf{R}^{n\times n}$ are bounded and Borel measurable. We assume that the function $\theta(x,u,t)$ is continuous in u, the derivatives $\partial f(x,t)/\partial x$, $\partial\beta(x,t)/\partial x$, $\theta(x,u,t)/\partial x$ are bounded, and there exists a number $\delta > 0$ such that $b(x,t) = \frac{1}{2}\beta(x,t)\,\beta(x,t)^\top \geq \delta I_n > 0$ ($\forall x,t$), where I_n is the unit matrix in $\mathbf{R}^{n\times n}$. Also, we assume that $\Pi(\mathbf{R}^m) < +\infty$.

Under these assumptions, equation (1) has the unique strong solution (see e.g. Theorem 2 of Gikhman and Skorohod (1972), p. 242).

Let a bounded hypersurface $\Gamma(t)$ of $n-1$ dimension be given for almost all $t \in [0,T]$, and let $\partial\Gamma(t)$ be its edge (it can happen that $\partial\Gamma(t) = \emptyset$).

Let some number $T > 0$ be given. Let $\mathbb{I}$ denotes the indicator function, and let $|\cdot|$ denotes the Euclidean norm.

We shall study the occupation time of $y(t)$ in $\Gamma(t)$. More precisely, we shall study the limit of the random variables

$$l_\varepsilon(T) \triangleq \frac{1}{\varepsilon}\int_0^T \mathbb{I}\{y(t) \in \Gamma(\varepsilon,t)\}\,dt \tag{2}$$

as $\varepsilon \to 0+$, where

$$\Gamma(\varepsilon,t) \triangleq \left\{x \in \mathbf{R}^n : \inf_{y\in\Gamma(t)} |x-y| < \frac{\varepsilon}{2}\right\}. \tag{3}$$

Spaces and Classes of Functions

We denote by $\|\cdot\|_X$ the norm in a normed space X, and we denote by $(\cdot,\cdot)_X$ the scalar product in a Hilbert space X.

We introduce some spaces of (complex valued) functions. Let $H^0 \triangleq L_2(\mathbf{R}^n)$, $H^1 \triangleq W_2^1(\mathbf{R}^n)$, where $W_q^m(\mathbf{R}^n)$ is the Sobolev space of functions which belong $L_q(\mathbf{R}^n)$ together with first m derivatives, $q \geq 1$.

Let H^{-1} be the dual space to H^1, with the norm $\|\cdot\|_{H^{-1}}$ such that $\|u\|_{H^{-1}}$ for $u \in H^0$ is the supremum of $(u,v)_{H^0}$ over all $v \in H^0$: $\|v\|_{H^1} \leq 1$.

Let ℓ_m denotes the Lebesgue measure in $\mathbf{R}^m$, let $\overline{\mathcal{B}}_m$ be the σ–algebra of the Lebesgue sets in $\mathbf{R}^m$. We introduce the following spaces:

$$C^0 \triangleq C([0,T];H^0), \quad X^k \triangleq L^2([0,T],\overline{\mathcal{B}}_1,\ell_1;H^k), \quad k = 0,\pm 1,$$

and $Y^1 \triangleq X^1 \cap C^0$, with the norm $\|u\|_{Y^1} \triangleq \|u\|_{X^1} + \|u\|_{C^0}$.

The scalar product $(u,v)_{H^0}$ is assumed to be well defined for $u \in H^{-1}$ and $v \in H^1$ as well (extending it in a natural manner from $u \in H^0$ and $v \in H^1$).

Let μ be a real number such that

$$\mu \in \begin{cases} (1,2) & \text{if } n = 1 \\ \left(1, n(n-1)^{-1}\right) & \text{if } n > 1. \end{cases}$$

We introduce the space $\mathcal{W} = W_\mu^1(\mathbf{R}^n)$ with the norm

$$\|u\|_{\mathcal{W}} \triangleq \|u\|_{L_\mu(\mathbf{R}^n)} + \textstyle\sum_{i=1}^n \|\frac{\partial u}{\partial x_i}\|_{L_\mu(\mathbf{R}^n)}$$

and its conjugate space $\mathcal{W}^*$, as well as the space $\mathcal{X} \triangleq L^\infty([0,T],\bar{\mathcal{B}}_1,\ell_1;\mathcal{W}^*)$.

Proposition 2.1 *If $\xi \in H^1$, $\eta \in H^0$ then $\xi\eta \in \mathcal{W}$ and $\|\xi\eta\|_{\mathcal{W}} \le c\|\xi\|_{H^1}\|\eta\|_{H^0}$, where $c = c(n,D,\mu)$ is a constant.*

Proof. For $r = 2/\mu$, $r' = r(r-1)^{-1}$, $p = \mu r'$ we have that

$$\|\xi\eta\|_{\mathcal{W}} = \left(\int_D \xi^\mu \eta^\mu dx\right)^{1/\mu} \le \left(\left(\int_D \xi^{\mu r'} dx\right)^{1/r'} \left(\int_D \eta^2 dx\right)^{1/r}\right)^{1/\mu}$$
$$= \|\eta\|_{H^0} \left(\int_D \xi^p dx\right)^{1/p}. \tag{4}$$

Furthermore, $p = \mu r' = 2\mu(2-\mu)^{-1}$, $\mu(2-\mu)^{-1} < n(n-2)^{-1}$ and $p < 2n(n-2)^{-1}$ for $n > 2$. Using standard embedding theorems for the Sobolev spaces (see e.g. [17]), we obtain the estimate $\|\xi\|_{L_p(D)} \le c\|\xi\|_{H^1}$ for a constant $c = c(n,D,\mu)$. This completes the proof of Proposition 2.1.□

Proposition 2.2 *If $\xi \in H^1$, $g \in \mathcal{W}^* \cap H^0$ then $\xi g \in H^{-1}$ and $\|\xi g\|_{H^{-1}} \le c\|\xi\|_{H^1}\|g\|_{\mathcal{W}^*}$ for a constant $c = c(n,D,\mu)$.*

Proof. Introduce the set $BH^1 = \{\eta \in H^1 \cap L_\infty(D) : \|\eta\|_{H^1} \le 1\}$. We have that

$$\|\xi g\|_{H^{-1}} = \sup_{\eta \in BH^1} (\eta, \xi g)_{H^0} \le \sup_{\eta \in BH^1} (\eta\xi, g)_{H^0} \le \sup_{\eta \in BH^1} \|\eta\xi\|_{\mathcal{W}}\|g\|_{\mathcal{W}^*}$$
$$\le c \sup_{\eta \in BH^1} \|\eta\|_{H^1}\|\xi\|_{H^1}\|g\|_{\mathcal{W}^*} \le c\|\xi\|_{H^1}\|g\|_{\mathcal{W}^*}, \tag{5}$$

where c is the constant from Proposition 2.1. This completes the proof of Proposition 2.2.□

3 On Solvability of Integro-differential Kolmogorov-Fokker-Planck Equations

In this section an integro-differential analog of the Kolmogorov–Fokker–Planck equation is studied. Some existence results theorems for these equations can be found in Carrany and Menaldy (1992). However, we shall need to extend the existence for a case when there are generalized functions in coefficients.

Let $\mathcal{A} = \mathcal{A}(t)$ be the integro-differential operator generated by the process $y(t)$:

$$\mathcal{A} \stackrel{\Delta}{=} \mathcal{A}_c + \mathcal{J},$$

where

$$\mathcal{A}_c v = \mathcal{A}_c(t) v \stackrel{\Delta}{=} \sum_{i,j=1}^n b_{ij}(x,t) \frac{\partial^2 v}{\partial x_i \partial x_j}(x) + \sum_{i=1}^n \widehat{f}_i(x,t) \frac{\partial v}{\partial x_i}(x), \tag{6}$$

$$\mathcal{J} v \stackrel{\Delta}{=} \mathcal{J}'(t) v - \Pi(R^m) v, \quad \mathcal{J}'(t) v \stackrel{\Delta}{=} \int_{\mathbf{R}^m} v(x + \theta(x,u,t)) \Pi(du), \tag{7}$$

and where

$$\widehat{f}(x,t) \stackrel{\Delta}{=} f(x,t) - \int_{\mathbf{R}^m} \theta(x,u,t) \Pi(du).$$

Here $b_{ij}, \widehat{f}_i, x_i$ are the components of the matrix b and the vectors $\widehat{f}, x$.

Let $Q \triangleq \mathbf{R}^n \times [0,T]$. We consider a boundary value problem in Q

$$\begin{cases} \frac{\partial V}{\partial t} + \mathcal{A}V + gV = -\varphi, \\ V(x,T) = R(x). \end{cases} \tag{8}$$

As is known, this problem (8) is uniquely solvable in the class Y^1 for $\varphi \in X^{-1}$, $R \in H^0$ and $g \in L_\infty(Q)$ (see e.g. Ladyzenskaya *et al* (1968)).

Condition 3.1 *There exists a constant $c_\pi > 0$ such that*

$$\mathbf{P}(\eta \in B) \le c_\pi \ell_n(B) \ell_n(D)^{-1}$$

for any $t \in [0,T]$, any bounded measurable set $D \subset \mathbf{R}^n$ and any measurable set $B \subseteq D$, where $\eta \triangleq \eta_1 + \theta(\eta_1, \eta_2, t)$, and where $\eta_1 : \Omega \to \mathbf{R}^n$ and $\eta_2 : \Omega \to \mathbf{R}^m$ are independent random vectors such that η_2 has the distribution described by the measure $\Pi(\cdot)/\Pi(\mathbf{R}^m)$, and $\mathbf{P}(\eta_1 \in B) = \ell_n(B)\ell_n(D)^{-1}$.

Notice that Condition 3.1 is satisfied if, for example, $\theta(x,u,t) \equiv \theta(u,t)$, i.e. does not depend on x. Another example is described in the following proposition.

Proposition 3.1 *Let there exist an integer $K > 0$, $p_1, ..., p_K \in \mathbf{R}$ and $u_1, ..., u_K \in \mathbf{R}^m$ such that*

$$\int_{\mathbf{R}^m} \xi(u)\Pi(du) = \sum_{i=1}^{K} \xi(u_i) p_i$$

for any $\xi(\cdot) \in C(\mathbf{R}^m)$. Set $F_i(x,t) \triangleq x + \theta(x, u_i, t)$, $D_i \triangleq F_i(\mathbf{R}^n, t)$. Let there exists the inverse function $F^{-1}(x,t) : D_i \to \mathbf{R}^n$ for any given t (i.e $F^{-1}(F(x,t),t) \equiv x$). Moreover, let there exists a conctant $c > 0$ such that $\ell_n(F_i^{-1}(B,t)) \le c\ell_n(B)$ for any measurable $B \subset D_i$, $i = 1, 2, ..., n$. Then Condition 3.1 is satisfied.

Proof. Let $D \subset \mathbf{R}^n$, η and η_1 be such as in Condition 3.1. For any measurable $B \subset D$, we have

$$\mathbf{P}(\eta \in B) = \sum_{i=1}^{n} p_i \mathbf{P}(F_i(\eta_1) \in B) = \sum_{i=1}^{n} p_i \mathbf{P}(\eta_1 \in F_i^{-1}(B)) = \frac{\ell_n(F_i^{-1}(B))}{\ell_n(D)} \le c \frac{\ell_n(B)}{\ell_n(D)}.$$

This completes the proof. □

We introduce the following parameter:

$$\mathcal{P} \triangleq \left(n, T, \delta, \sup_{x,t} |f(x,t)|, \sup_{x,t} |\beta(x,t)|, \sup_{x,t,i} \left| \frac{\partial \beta(x,t)}{\partial x_i} \right|, \Pi(\mathbf{R}^m), c_\pi \right). \tag{9}$$

Theorem 3.1 *Let Condition 3.1 be satisfied. Let $g \in X$, $\varphi \in X^{-1}$ and $R \in H^0$ be given. Let $g_\varepsilon \in L_\infty(Q) \cap X$ be such that $\|g_\varepsilon - g\|_X \to 0$ as $\varepsilon \to 0+$. Then*

(i) For any $\varepsilon > 0$, there exist the unique solution $V = V_\varepsilon \in Y^1$ of problem (8) with $g = g_\varepsilon$.

(ii) *The sequence V_ε has a limit V in Y^1 as $\varepsilon \to 0+$. This limit is uniquely defined by φ, g, and V does not depend on the sequences $\{g_\varepsilon\}$. Moreover, $gV \in X^{-1}$ and*

$$\|V\|_{Y^1} \le c\left(\|\varphi\|_{X^{-1}} + \|R\|_{H^0}\right), \tag{10}$$

where $c > 0$ is a constant which depends only on the parameters $\mathcal{P}$, μ and $\|g\|_X$.

Remark 3.1 *It can be seen from the proof that Theorem 3.1 holds even if derivatives $\partial f(x,t)/\partial x$ and $\partial\theta(x,t)/\partial x$ do not exist.*

Definition 3.1 *The limit V defined in Theorem 3.1 is said to be the solution in Y^1 of the problem (8) with $g \in X$, $\varphi \in X^{-1}$, and $R \in H^0$.*

Note that V depends linearly on (φ, R) for a given g. Moreover, by (10), it follows that $V = 0$ if $\varphi = 0$ and $R = 0$. Hence it follows that the operator assigning the solution V to the pair $(\varphi, R) \in X^{-1} \times H^0$ is also linear and homogeneous.

Definition 3.2 *For every $g \in X$, define linear continuous operators $L(g) : Y^{-1} \to Y^1$ and $\mathcal{L}(g) : H^0 \to Y^1$ such that $V = L(g)R$ for V which is the solution in Y^1 of problem (8) with $g \in X$, $\varphi = 0$ and $R \in H^0$, and $V = \mathcal{L}(g)R$ for V which is the solution in Y^1 of problem (8) with $g \in X$, $\varphi \in X^{-1}$ and $R = 0$.*

The fact that these operators are continuous follows immediately from Theorem 3.1. Clearly, $V = L(g)\varphi + \mathcal{L}(g)R$ for V which is the solution in Y^1 of problem (8) with $g \in X$, $\varphi \in X^{-1}$ and $R \in H^0$.

To prove Theorem 3.1, we need first a preliminary lemma.

Lemma 3.1 *Let $\varepsilon > 0$ be such that there exist a solution $V = V_\varepsilon \in Y_1$ of problem (8) with $g = g_\varepsilon$. Then*

$$\|V_\varepsilon\|_{Y^1} \le c\left(\|\varphi\|_{X^{-1}} + \|R\|_{H^0}\right), \tag{11}$$

where $c > 0$ is a constant which depends only on the parameters $\mathcal{P}$, μ and $\|g\|_X$.

Proof of Lemma 3.1. We shall use below the elementary estimate $uv \le u^2/(2\gamma) + v^2\gamma/2$ $(\forall u, v, \gamma \in \mathbf{R}, \gamma > 0)$.

Let $v \in H^1 \cap C^2(\mathbf{R}^n)$. For $t \in [0,T]$, we have that

$$\begin{aligned}
(v, \mathcal{A}_c(t)v)_{H^0} &= \left(v, \sum_{i,j=1}^n b_{ij}\frac{\partial^2 v}{\partial x_i \partial x_j}\right)_{H^0} + \left(v, \sum_{i=1}^n \widehat{f}_i \frac{\partial v}{\partial x_i}\right)_{H^0} \\
&= -\sum_{i,j=1}^n \left(\frac{\partial v}{\partial x_i}, b_{ij}\frac{\partial v}{\partial x_j}\right)_{H^0} - \sum_{i,j=1}^n \left(v, \frac{\partial b_{ij}}{\partial x_i}\frac{\partial v}{\partial x_j}\right)_{H^0} + \sum_{i=1}^n \left(v, \widehat{f}_i \frac{\partial v}{\partial x_i}\right)_{H^0} \\
&\le -\delta \sum_{i=1}^n \left\|\frac{\partial v}{\partial x_i}\right\|_{H^0}^2 + \sum_{i,j=1}^n \|v\|_{H^0} \left\|\frac{\partial b_{ij}}{\partial x_i}\right\|_{L_\infty(Q)} \left\|\frac{\partial v}{\partial x_j}\right\|_{H^0} \\
&+ \sum_{i=1}^n \|v\|_{H^0} \|\widehat{f}_i\|_{L_\infty(Q)} \left\|\frac{\partial v}{\partial x_i}\right\|_{H^0} \\
&\le -\delta \sum_{i=1}^n \left\|\frac{\partial v}{\partial x_i}\right\|_{H^0}^2 + \frac{\delta}{4}\sum_{i=1}^n \left\|\frac{\partial v}{\partial x_i}\right\|_{H^0}^2 \\
&+ \frac{C}{\delta}\|v\|_{H^0}^2 \sum_{i,j=1}^n \left(\left\|\frac{\partial b_{ij}}{\partial x_i}\right\|_{L_\infty(Q)}^2 + \|f_i\|_{L_\infty(Q)}^2\right),
\end{aligned} \tag{12}$$

where $C = C(n)$ is a constant. Hence we obtain the inequality

$$(v, \mathcal{A}_c(t)v)_{H^0} \leq -\frac{3\delta}{4}\sum_{i=1}^{n}\left\|\frac{\partial v}{\partial x_i}\right\|_{H^0}^2 + C_1'\|v\|_{H^0}^2$$

for all $v \in H^1$, $t \in [0,T]$, where a constant C_1' depends only on $\mathcal{P}$.

Let $D_K \triangleq \{x \in \mathbf{R}^n : |x| \leq K\}$. By Condition 3.1,

$$\begin{aligned}
\|\mathcal{J}'(t)v\|_{H^0}^2 &= \int_{\mathbf{R}^n} dx \left|\int_{\mathbf{R}^m} v(x+\theta(t,x,u))\Pi(du)\right|^2 \\
&\leq \int_{\mathbf{R}^n} dx \int_{\mathbf{R}^m} |v(x+\theta(t,x,u))|^2 \Pi(du) \\
&= \lim_{K\to+\infty} \int_{D_K} dx \int_{\mathbf{R}^m} |v(x+\theta(t,x,u))|^2 \Pi(du) \\
&= \lim_{K\to+\infty} \ell_n(D_K)\int_{\mathbf{R}^n} \mu_K(dy)|v(y)|^2 \leq c_\pi \int_{\mathbf{R}^n} |v(y)|^2 dy = c_\pi \|v\|_{H^0}^2,
\end{aligned}$$

where $\mu_K(\cdot)$ is the probability measure which describes the distribution of a random vector $\eta = \eta_K$ such as in Condition 3.1, where $D = D_K$. Then

$$\|\mathcal{J}(t)v\|_{H^0}^2 \leq (c_\pi + \Pi(\mathbf{R}^m))\|v\|_{H^0}^2. \tag{13}$$

Thus,

$$(v, \mathcal{A}(t)v)_{H^0} \leq -\frac{3\delta}{4}\sum_{i=1}^{n}\left\|\frac{\partial v}{\partial x_i}\right\|_{H^0}^2 + C_1\|v\|_{H^0}^2 \tag{14}$$

for all $v \in H^1$ and $t \in [0,T]$, where a constant C_1' depends only on $\mathcal{P}$.

Furthermore, we have that

$$(v, \varphi_\varepsilon(\cdot,t))_{H^0} \leq \|v\|_{H^1}\|\varphi_\varepsilon(\cdot,t)\|_{H^{-1}} \leq \frac{\delta}{4}\left(\sum_{i=1}^{n}\left\|\frac{\partial v}{\partial x_i}\right\|_{H^0}^2 + \|v\|_{H^0}^2\right) + C_2\|\varphi_\varepsilon(\cdot,t)\|_{H^{-1}}^2 \tag{15}$$

for all $v \in H^1$ and $t \in [0,T]$, where a constant C_2 also depends only on $\mathcal{P}$.

Proposition 2.1 yields

$$\begin{aligned}
(v, g_\varepsilon v)_{H^0} &\leq \|v^2\|_{\mathcal{W}}\|g_\varepsilon\|_{\mathcal{W}^*} \leq C_3\|v\|_{H^1}\|v\|_{H^0}\|g_\varepsilon\|_{\mathcal{W}^*} \\
&\leq \tfrac{\delta}{4}\textstyle\sum_{i=1}^{n}\|v\|_{H^1}^2 + \widehat{C}_3\|v\|_{H_0}^2 \leq \tfrac{\delta}{4}\textstyle\sum_{i=1}^{n}\left\|\frac{\partial v}{\partial x_i}\right\|_{H^0}^2 + C_3\|v\|_{H_0}^2 \quad \forall v \in H^1,
\end{aligned} \tag{16}$$

where constants $\widehat{C}_3$ and C_3 depend on $\|g_\varepsilon\|_{\mathcal{W}^*}$, δ and n.

For the solution $V = V_\varepsilon$ of problem (8) with $g = g_\varepsilon$, $\varepsilon \in (0, \varepsilon_1]$, we have from (12)- (16) that

$$\begin{aligned}
\|V_\varepsilon(\cdot,t)\|_{H^0}^2 \quad -\|V_\varepsilon(\cdot,T)\|_{H^0}^2 &= 2\int_t^T (V_\varepsilon(\cdot,s), \mathcal{A}V_\varepsilon(\cdot,s) + g_\varepsilon V_\varepsilon(\cdot,s) + \varphi(\cdot,s))_{H^0} ds \\
&\leq \int_t^T \left\{-\delta\textstyle\sum_{i=1}^{n}\left\|\frac{\partial V_\varepsilon}{\partial x_i}(\cdot,s)\right\|_{H^0}^2 + C_4(\|V_\varepsilon(\cdot,s)\|_{H^0}^2 + \|\varphi(\cdot,s)\|_{H^{-1}}^2)\right\} ds.
\end{aligned} \tag{17}$$

where a constant C_4 depends on $\mathcal{P}$, μ and $\|g\|_X$. Thus,

$$\|V_\varepsilon\|_{Y^1} \le C_*(\|\varphi\|_{X^{-1}} + \|R\|_{H^0}) \quad \forall \varepsilon \in (0, \varepsilon_1], \tag{18}$$

where a constant C_* also depends on $\mathcal{P}$, μ and $\|g\|_X$. This completes the proof of Lemma 3.1. □

Proof of Theorem 3.1. Let us prove (i). Consider a boundary value problem

$$\begin{cases} \frac{\partial V}{\partial t} + \mathcal{A}_c V + gV = -\varphi, \\ V(x,T) = R(x), \end{cases} \tag{19}$$

where $g \in L_\infty(Q)$. The solution $V = V_\varepsilon \in Y^1$ of this problem with $g = g_\varepsilon$ is well defined. For every $g \in L_\infty(Q)$, introduce linear continuous operators $L_c(g) : X^{-1} \to Y^1$ and $\mathcal{L}_c(g) : H^0 \to Y^1$ such that $V = L_c(g)\varphi + \mathcal{L}_c(g)R$ for V which is the solution in Y^1 of problem (19) with given g, φ and R.

The solution $V = V_\varepsilon$ of (8) with $g = g_\varepsilon$ (if exists) has the form $V = L_c(g_\varepsilon)\varphi + \mathcal{L}_c(g_\varepsilon)R + L_c(g_\varepsilon)\mathcal{J}V$. Let

$$\begin{aligned} &V_0 \triangleq 0 \in Y^1, \\ &V_k \triangleq L_c(g)\varphi + \mathcal{L}_c(g)R + L_c(g)\mathcal{J}V_{k-1}, \quad U_k \triangleq V_k - V_{k-1}, \quad k = 1,2,\dots . \end{aligned}$$

It suffices to prove that $U_k \to 0$ in Y^1 as $k \to +\infty$. Set

$$y_k(t) \triangleq \|U_k(\cdot,t)\|_{H^0} + \delta \sum_{i=1}^n \int_t^T \left\| \frac{\partial U_k}{\partial x_i}(\cdot,s) \right\|_{H^0}^2 ds.$$

Similarly (17), we have

$$y_k(t) \le c_1 + C_4 \int_t^T (y_k(s) + y_{k-1}(s))ds.$$

where $c_1 \triangleq \|R(\cdot)\|_{H^0}$. By the Bellman inequality,

$$y_k(t) \le c_1 e^{C_4(T-t)} \int_t^T y_{k-1}(s)ds.$$

It easy to see that $y_k(t) \le C^k$, where $C > 0$ is a constant independent on k and t. After standard iterations, we have that $\sup_{t\in[0,T]} y_k(t) \to 0$ as $k \to +\infty$. Thus, $\{V_k\}$ is a Cauchy sequence in Y^1. Then it has the limit V_ε, and this limit is a solution. The uniqueness follows from a prior estimate from Lemma 3.1. Then (i) follows.

Let us show that the sequence $\{V_\varepsilon\}$, $\varepsilon \to 0$, is a Cauchy sequence in the space Y^1. Let $\varepsilon_1 \to 0$, $\varepsilon_2 \to 0$. Let $W = V_{\varepsilon_1} - V_{\varepsilon_2}$, then

$$\begin{cases} \frac{\partial W}{\partial t} + \mathcal{A}W + g_{\varepsilon_1}W = -\xi, \\ W(x,T) = 0, \end{cases} \tag{20}$$

where $\xi \triangleq (g_{\varepsilon_1} - g_{\varepsilon_2})V_{\varepsilon_2}$. Furthermore,

$$\|g_{\varepsilon_1} - g_{\varepsilon_2}\|_{W^*} \to 0, \tag{21}$$

because $\{g_\varepsilon\}$ is a Cauchy sequences. By Proposition 2.2 and (18), (21),

$$\|(g_{\varepsilon_1}-g_{\varepsilon_2})V_{\varepsilon_2}\|_{X^{-1}} = \int_0^T \|(g_{\varepsilon_1}-g_{\varepsilon_2})V_{\varepsilon_2}(\cdot,t)\|_{H^{-1}}\,dt$$
$$\leq \int_0^T \|g_{\varepsilon_1}-g_{\varepsilon_2}\|_{\mathcal{W}^*}\|V_{\varepsilon_2}(\cdot,t)\|_{H^1}\,dt \leq \|g_{\varepsilon_1}-g_{\varepsilon_2}\|_X \int_0^T \|V_{\varepsilon_2}(\cdot,t)\|_{H^1}\,dt$$
$$= \|g_{\varepsilon_1}-g_{\varepsilon_2}\|_{\mathcal{W}^*}\|V_{\varepsilon_2}\|_{X^1} \to 0 \qquad (22)$$

as $\varepsilon_1 \to 0$, $\varepsilon_2 \to 0$. Hence $\|\xi\|_{X^{-1}} \to 0$. Estimate (18) applied to the solution W of boundary value problem (20) yields

$$\|W\|_{Y^1} \leq C_*\|\xi\|_{X^{-1}} \to 0.$$

Hence the sequence $\{V_\varepsilon\}$, $\varepsilon \to 0$, is a Cauchy sequence (and has a limit) in the Banach space Y^1. Estimate (10) and the uniqueness of V follows from (18),(20). This completes the proof of Theorem 3.1. □

Corollary 3.1 *Let* $V_\varepsilon \triangleq L(g_\varepsilon)\varphi_\varepsilon + \mathcal{L}(g_\varepsilon)R_\varepsilon$, *and* $V \triangleq L(g)\varphi + \mathcal{L}(g)R$, *where* $g, g_\varepsilon \in X$, $\varphi, \varphi_\varepsilon \in X^{-1}$ *are such that* $\|g_\varepsilon - g\|_X \to 0$, $\|\varphi_\varepsilon - \varphi\|_{X^{-1}} \to 0$ *and* $\|R_\varepsilon - R\|_{H^0} \to 0$ *as* $\varepsilon \to 0+$. *Then* $\|V_\varepsilon - V\|_{Y^1} \to 0$.

Proof. For a sake of simplicity, assume that $\varphi_\varepsilon \equiv \varphi$. Let $\|\mathcal{L}(g_\varepsilon)\|$ denotes the norm of the operator $\mathcal{L}(g_\varepsilon): H^0 \to Y^1$. By Theorem 3.1, $\sup_\varepsilon \|\mathcal{L}(g_\varepsilon)\| \leq \text{const}$. Then

$$\begin{aligned}\|V_\varepsilon - V\|_{Y^1} &\leq \|\mathcal{L}(g_\varepsilon)R_\varepsilon - \mathcal{L}(g_\varepsilon)R\|_{Y^1} + \|\mathcal{L}(g_\varepsilon)R - \mathcal{L}(g)R\|_{Y^1}\\ &\leq \sup_\varepsilon \|\mathcal{L}(g_\varepsilon)\|\|R_\varepsilon - R\|_{H^0} + \|\mathcal{L}(g_\varepsilon)R - \mathcal{L}(g)R\|_{Y^1}.\end{aligned}$$

By Theorem 3.1, it also follows that $\|\mathcal{L}(g_\varepsilon)R - \mathcal{L}(g)R\|_{Y^1} \to 0$ as $\varepsilon \to 0$. Then the proof follows. □

4 Adjoint Equations

Let $\mathcal{A}_c^* = \mathcal{A}_c^*(t)$ be the operator which is formally adjoint to the operator $\mathcal{A}_c(t)$ defined by (6):

$$A_c^*(t)p = \sum_{i,j=1}^{n} \frac{\partial^2}{\partial x_i \partial x_j}\Big(b_{ij}(x,t)p(x)\Big) - \sum_{i=1}^{n} \frac{\partial}{\partial x_i}\Big(\widehat{f}_i(x,t)p(x)\Big). \qquad (23)$$

Let $\mathcal{J}^* = \mathcal{J}^*(t): H^0 \to H^0$ be the operator which is adjoint to the operator $\mathcal{J} = \mathcal{J}(t): H^0 \to H^0$ defined by (7). Let $\mathcal{A}^* \triangleq \mathcal{A}_c^* + \mathcal{J}^*$. Consider the following boundary value problem in Q:

$$\begin{cases} \frac{\partial p}{\partial t} = \mathcal{A}^* p + gp + \varphi, \\ p(x,0) = \rho(x). \end{cases} \qquad (24)$$

Theorem 4.1 *Assume that Condition 3.1 is satisfied. Let* $g \in X$, $\varphi \in X^{-1}$ *and* $\rho \in H^0$ *be given. Let* $g_\varepsilon \in L_\infty(Q) \cap X$ *be such that*

$$\|g_\varepsilon - g\|_X \to 0 \quad as \quad \varepsilon \to 0+.$$

Then:

(i) *For any* $\varepsilon > 0$, *there exist the unique solution* p_ε *of (24) with* $g = g_\varepsilon$.

(ii) *The sequence* p_ε *has an unique limit* p *in* Y^1 *as* $\varepsilon \to 0+$, *and*

$$\|p\|_{Y^1} \leq c(\|\varphi\|_{X^{-1}} + \|\rho\|_{H^0}),$$

where a constant $c > 0$ *depends only on the parameters* $\mathcal{P}$, μ *and* $\|g\|_X$.

The proof of Theorem 4.1 is similar to the proof of Theorem 4.2. Note only that, by Remark 3.1, it follows that the coefficients of the operator $\mathcal{A}_c^*$ are smooth enough, and, by (13), $\|\mathcal{I}^* v\|_{H^0}^2 \leq (c_\pi + \Pi(\mathbf{R}^m))\|v\|_{H^0}^2$ for all $v \in H^0$.

For $g \in X$, introduce a linear continuous operators $\widehat{\mathbf{L}}(g) : X^{-1} \to H^0$ and $\mathbf{L}_T(g) : H^0 \to H^0$ such that $V(\cdot,0) = \widehat{\mathbf{L}}(g)\varphi + \mathbf{L}_T(g)R$, where $V = L(g)\varphi + \mathcal{L}R$ is the solution of problem (8).

Proposition 4.1 *For* p, g, φ *and* ρ *from Theorem 4.1, we have that* $p = L(g)^*\varphi + \widehat{\mathbf{L}}(g)^*\rho$ *and* $p(\cdot,T) = L(g)^*\varphi + \mathbf{L}_T(g)^*\rho$, *where* $L(g)^* : X^{-1} \to X^1$, $\widehat{\mathbf{L}}(g)^* : H^0 \to X^1$ *and* $\mathbf{L}_T(g)^* : H^0 \to H^0$ *are linear continuous operators which are adjoint to the operators* $L(g) : X^{-1} \to X^1$ *and* $\widehat{\mathbf{L}}(g) : X^{-1} \to H^0$ *and* $\mathbf{L}_T(g) : H^0 \to H^0$ *correspondingly.*

Proof. Let $\phi \in X^0$, $R \in H^0$ be arbitrary, $V = L(g)\phi + \mathcal{L}(g)R$. Then

$$\begin{aligned}
(p(\cdot,T),R)_{H^0} - (\rho, \widehat{\mathbf{L}}(g)\phi + \mathbf{L}_T R)_{H^0} &= (p(\cdot,T), V(\cdot,T))_{H^0} - (p(\cdot,0), V(\cdot,0))_{H^0} \\
&= \left(\frac{\partial p}{\partial t}, V\right)_{X^0} + \left(p, \frac{\partial V}{\partial t}\right)_{X^0} \\
&= (\mathcal{A}^* p + gp + \varphi, V)_{X^0} + (p, -\mathcal{A}V - gV - \phi)_{X^0} \\
&= (\varphi, V)_{X^0} - (p,\phi)_{X^0} = (\varphi, L(g)\phi + \mathcal{L}(g)R)_{X^0} - (p,\phi)_{X^0}.
\end{aligned}$$

Then

$$(p(\cdot,T),R)_{H^0} + (p,\phi)_{X^0} = (\rho, \widehat{\mathbf{L}}(g)\phi + \mathbf{L}_T(g)R)_{H^0} + (\varphi, L(g)\phi + \mathcal{L}(g)R)_{X^0}.$$

Then the proof follows. □

Condition 4.1 *There exists uniformly bounded derivatives* $\partial^k \beta(x,u,t)/\partial x^k$, $\partial^k f(x,t)/\partial x^k$ *and* $\partial^k \theta(x,u,t)/\partial x^k$ *for* $k = 1,2$.

Theorem 4.2 *Let Conditions 3.1, 4.1 be satisfied, let* $g(x,t) : Q \to \mathbf{R}$ *be a Borel measurable function which belongs* X *and is bounded together with the derivatives* $\partial^k g(x,t)/dx^k$ *for* $k = 1,2$. *Let the vector* a *in (1) has the probability density function* $\rho \in H^0$, *and let* $p \triangleq \widehat{\mathbf{L}}(g)^*\rho$. *Then*

$$\mathbf{E}R(y(T))\exp\left(\int_0^T g(y(t),t)dt\right) = \int_{\mathbf{R}^n} p(x,T)R(x)dx \tag{25}$$

for all Borel measurable $R(\cdot) \in H^0$. *In particular, if* $g = 0$ *then* $p(x,t)$ *is the probability density function of the solution* $y(t)$ *of equation (1).*

Proof. It suffices to prove (25) with $R(\cdot) \in H^0 \cap C^2(\mathbf{R}^n)$. For $(x,s) \in Q$, set $V(x,s) \triangleq \mathbf{E}\{R(y(T))|y(s) = x\}$. By Theorem 4 of Gikhman and Skorohod (1972), p. 296, it follows that $V = \mathcal{L}(0)R$. By Proposition 4.1, it follows that

$$\mathbf{E}R(y(T)) = (V(\cdot,0),\rho)_{H^0} = (\mathbf{L}_T(g)R,\rho)_{H^0} = (R,\mathbf{L}_T(g)^*\rho)_{H^0} = (R,p(\cdot,T))_{H^0}$$

for all $R(\cdot) \in H^0 \cap C^2(\mathbf{R}^n)$. This completes the proof. □

Corollary 4.1 *Let $V = \mathcal{L}(g)R$, where $R \in H^0$. Then there exist a version of V such that* $\operatorname{ess\,sup}_{x,t} V(x,t) \leq \max_x R(x)$ *and* $\operatorname{ess\,sup}_{x,t} V(x,t) \leq \min_x R(x)$.

Proof. If $R(\cdot) \in H^0 \cap C^2(\mathbf{R}^n)$ then $V(x,s) = \mathbf{E}\{R(y(T))|y(s) = x\}$ and the proof follows. For the general case $R \in H^0$, the proof can be obtained by a standard approximation. □

Condition 4.2 *(i) There exists uniformly bounded derivatives $\partial^m\beta(x,u,t)/\partial x^m$ for $m \leq 4$, $\partial^l f(x,t)/\partial x^l$ for $l = 1,2,3$, and $\partial^k\theta(x,u,t)/\partial x^k$ for $k = 1,2$.*

(ii) There exist $c_* \in \mathbf{R}$, a measure $\Pi_*(\cdot)$ in $\mathbf{R}^m$ and a bounded and Borel measurable function $\theta_*(x,u,t) : \mathbf{R}^n \times \mathbf{R}^m \times \mathbf{R} \to \mathbf{R}^{n\times n}$ which is continuous in u, such that $\Pi_*(\mathbf{R}^m) < +\infty$ and $\mathcal{J}^*(t)v = \int_{\mathbf{R}^m} v(x+\theta_*(x,u,t))\Pi_*(du) + c_*v$ for any $v \in H^0$.

(iii) The derivatives $\partial^k\theta_*(x,u,t)/\partial x^k$ are bounded for $k = 1,2$ and Condition 3.1 is satisfied with substituting $(\Pi(\cdot),\theta(\cdot)) = (\Pi_*(\cdot),\theta_*(\cdot))$.

Note that if the mapping $z = x+\theta(x,u,t)$ maps $\mathbf{R}^n$ one-to-one onto itself for any (u,t), then $\theta_*(\cdot)$ can be found such that $x = z-\theta_*(z,u,t)$ is the inverse mapping. If the last one is differentiable, then $\Pi_*(\cdot)$ can be found as $\Pi_*(dx) = J(x)\Pi(dx)$, where $J(x)$ is the Jacobian of the transformation $y = x-\theta_*(x,u,t)$ (see Gikhman and Skorohod (1972), p. 299).

Corollary 4.2 *Let Conditions 3.1, 4.1 and 4.2 be satisfied. Let $\rho \in L_\infty(\mathbf{R}^n) \cap H^0$, and let $p \triangleq \widehat{\mathbf{L}}(0)^*\rho$. Then $p \in L_\infty(Q)$.*

Proof. It can be seen that equations (25) after a change of time variable can be rewritten in form (6), and then Corollary 4.1 is satisfied. This completes the proof.□

5 On a Class of Acceptable Hupersurfaces

We shall use below the equations from Sections 3-4 for the distributions of the occupation time on hypersurface. In this Section we describe a class of acceptable hypersurfaces.

Let $\Gamma \subset D$ be some $(n-1)$-dimensional hypersurface.

We denote by $e^{(i)}$ the ith unit vectors in $\mathbf{R}^n$, $i = 1,...,n$.

Let $\mathbf{n}(x)$ be the normal to Γ in x, and let $\alpha_i(x)$ be the angle between $e^{(i)}$ and $\mathbf{n}(x)$.

Introduce functions $\gamma_i : \Gamma \to \mathbf{R}$, $i = 1,...,n$, such that

$$\gamma_i(x) = \begin{cases} |\cos\alpha_i(x)| & \text{if the normal } \mathbf{n}(x) \text{ at } x \text{ is uniquely defined} \\ 0 & \text{if the normal at } x \text{ is not defined.} \end{cases}$$

(In fact, $\mathbf{n}(x)$ is not defined in points of violation of smoothness of Γ).

Denote by $N(x,j,\Gamma)$ the number of intersections of the hypersurface Γ by the ray from $x=(x_1,x_2,...,x_n)$ to $(x_1,...,x_{j-1},-\infty,x_{j+1},...,x_n)$. Let $\widehat{x}_k(x,j)$ be the corresponding intersections points.

We assume that $N(x,j,\Gamma)=+\infty$, if the ray is tangential to Γ.

Set

$$G_j(x)\stackrel{\Delta}{=}\sum_{k=1}^{N(x,j,\Gamma)}\gamma_j\left(\widehat{x}_k(x,j)\right),\quad g=\sum_{j=1}^n\frac{\partial G_j}{\partial x_j},$$
$$\Gamma(\varepsilon)\stackrel{\Delta}{=}\{x\in\mathbf{R}^n:\ \inf_{y\in\Gamma}|x-y|\le\tfrac{\varepsilon}{2}\},\quad g_\varepsilon(x)\stackrel{\Delta}{=}\tfrac{1}{\varepsilon}\mathbb{I}\{x\in\Gamma(\varepsilon)\}.$$

Definition 5.1 *A set $\widehat{\Gamma}\in\mathbf{R}^n$ is said to be an $(n-1)$–dimensional polyhedron if there exist an integer N and $c_i\in\mathbf{R}^n$, $\delta_i\in\mathbf{R}$, $i=0,1,...,N$ such that $\widehat{\Gamma}=\{x\in\mathbf{R}^n:\ c_0'x=\delta_0,\ c_i'x\le\delta_i,\ i=1,...,N\}$. The set $\{x\in\mathbf{R}^n:\ c_0'x=\delta_0,\ c_i'x<\delta_i,\ i=1,...,N\}$ is said to be the interior of $\widehat{\Gamma}$.*

Lemma 5.1 *[Dokuchaev (1999)] Let a hypersurface $\Gamma\subset\mathbf{R}^n$ be bounded and such that there exists a set $\widehat{\Gamma}\subset\mathbf{R}^n$ and a continuous biection $\mathcal{M}:\mathbf{R}^n\to\mathbf{R}^n$ which satisfy the following assumptions:*

(i) $\Gamma=\mathcal{M}(\widehat{\Gamma})$;

(ii) $\widehat{\Gamma}=\cup_{i=1}^{\mathcal{N}}\widehat{\Gamma}_i$, *where $\mathcal{N}$ is an integer, $\widehat{\Gamma}_i$ is $(n-1)$–dimensional polyhedron;*

(iii) $\mathcal{M}:\widehat{\Gamma}_i\to\mathbf{R}^n$ *are C^1–smooth biections, $i=1,...,\mathcal{N}$;*

(iv) $|\mathbf{n}(x)-\mathbf{n}_i|\le\delta_0$, *if $\mathcal{M}^{-1}(x)$ belongs the interior of $\widehat{\Gamma}_i$, $i=1,...,\mathcal{N}$, where $x\in\Gamma$, $\delta_0\le n^{-2}/2$ is a constant, $\mathbf{n}(x)$ is the normal to Γ in x, and $\mathbf{n}_i$ is the normal to $\widehat{\Gamma}_i$; it is assumed that the orientations of these normals are fixed and $|\mathbf{n}(x)|=1,|\mathbf{n}_i|=1$;*

(v) $\mathcal{M}(x)=x$, *if x is a top point of some $\widehat{\Gamma}_i$.*

Then $N(j,x,\Gamma)<+\infty$ for a.e. x. Moreover, $g\in\mathcal{W}^*\cap H^{-1}$ and $g_\varepsilon(\cdot)\to g$ in $\mathcal{W}^*$.

Proof of Lemma 5.1. Let $\Gamma_i=\mathcal{M}(\widehat{\Gamma}_i)$. Introduce the set $S_0=\cup\partial\Gamma_i$; we have that $\ell_n(S_0)=0$. Consider a set of open regions $D_i\subseteq D$ such that $\bar{D}=\cup_{i=1}^{\mathcal{N}}\bar{D}_i$, $D_i\cap D_j=\emptyset$, if $i\ne j$, $\Gamma_i=\Gamma\cap\bar{D}_i$; we denote by $\bar{D}_i$, $\bar{D}$ closures of regions. Let $\Gamma_i(\varepsilon)=\Gamma(\varepsilon)\cap\bar{D}_i$. Let $P_j(x)$ be the straight line which crosses x and is parallel to $e^{(j)}$.

Introduce functions $\psi(.,\varepsilon):\Gamma(\varepsilon)\to\Gamma\backslash S_0$ such that $|\psi(x,\varepsilon)-x|\le 2\varepsilon$, and if $x\in\Gamma_i(\varepsilon)$ then $\psi(x,\varepsilon)\in\Gamma_i$.

Furthermore, introduce functions $G_j^\varepsilon(x)$, $x=(x_1,...,x_n)$ such that they are continuous by x_j, $G_j^\varepsilon(x)=0$, if $(x_1,...,x_{j-1},y_j,x_{j+1},...,x_n)\notin\Gamma(\varepsilon)$ $(\forall y_j<x_j)$ and $\partial G_j^\varepsilon(x)/\partial x_j=\varepsilon^{-1}\gamma_j^2(\psi(x,\varepsilon))$, $x\in P_j(x)\cap\Gamma(\varepsilon)$. We suppose also that the functions $G_j^\varepsilon(x)$ are constant at any segment of $P_j(x)$ which belongs $D\backslash\Gamma(\varepsilon)$. Of course, values of the constants are different at different segments. These conditions define $G_j^\varepsilon(x)$ uniquely.

It can be easily seen that $\sum_{j=1}^n\gamma_j^2(x)=1$ $(\forall x\in\Gamma\backslash S_0)$. Then

$$\sum_{j=1}^n\frac{\partial G_j^\varepsilon}{\partial x_j}(x)=\frac{1}{\varepsilon}\sum_{j=1}^n\gamma_j^2(\psi(x,\varepsilon))=\frac{1}{\varepsilon},\ x\in\Gamma(\varepsilon),\quad \sum_{j=1}^n\frac{\partial G_j^\varepsilon}{\partial x_j}(x)=0,\ x\notin\Gamma(\varepsilon),$$

$$\sum_{j=1}^{n} \frac{\partial G_j^{\varepsilon}}{\partial x_j}(x) = g_{\varepsilon}(x), \quad x \in D.$$

We have that $\ell_n(x \in \mathbf{R}^n : P_j(x) \cap S_0 \neq \emptyset) = 0$, $j = 1, ..., n$.

The fact that $N(j,x) < +\infty$ for a.e. x seems to be obvious; moreover, one also is a consequence from Proposition 2.1 [19]. Because of piecewise C^1–smoothness of Γ, we have that $N(x,j) \leq$ const for a.e. x.

Let $\mathcal{N} = 1$, $\Gamma = \Gamma_1$ be such that $\mathbf{n}_1 = (\mathbf{n}_1^{(1)}, ..., \mathbf{n}_1^{(n)})$, $|\mathbf{n}_1^{(j)}| = n^{-1/2}$, $j = 1,..,n$, then $\gamma_j(x) \geq \delta_1$ $(\forall x \in \Gamma \backslash S_0, j = 1, ..., n)$ for some $\delta_1 > 0$. Denote by $\mu_j(x,\varepsilon)$ the length of the minimal segment belonging $\Gamma(\varepsilon)$ with endpoints at $\partial\Gamma(\varepsilon)$ containing x and which is parallel to $e^{(j)}$. We have that $\varepsilon^{-1}\mu_j(x,\varepsilon) \to \gamma_j^{-1}(x)$, $G_j^{\varepsilon}(x) \to G_j(x)$ as $\varepsilon \to 0+$ for a.e. x, $(\forall j = \overline{1,n})$ because of the continuous dependence on the normal by a point at the C^1–smooth surface $\Gamma = \Gamma_1$. Then $\sup_x |G_j^{\varepsilon}(x)| \leq \varepsilon^{-1}\mu_j(x,\varepsilon)\sup_x \gamma_j^2(\psi(x,\varepsilon)) \leq c$, where $c = c(\Gamma)$ is a constant, $x \in P_j(x) \cap \Gamma(\varepsilon)$. Hence $\|G_j^{\varepsilon} - G_j\|_{L_{\nu}(\Gamma(1))} \to 0$ as $\varepsilon \to 0+$ $(\forall \nu > 1, j = \overline{1,n})$.

Furthermore, let $u \in C^2(D) \cap \mathcal{W} \cap H^1$. We have that

$$\begin{aligned} \langle g_{\varepsilon} - g, u\rangle &= \textstyle\sum_{j=1}^{n} \int_{\partial D}\Big(G_j^{\varepsilon}(x) - G_j(x)\Big)u(x)\cos(\mathbf{n}(x), e_j)dx \\ &- \textstyle\sum_{j=1}^{n} \int_D \Big(G_j^{\varepsilon}(x) - G_j(x)\Big)\frac{\partial u}{\partial x_j}(x)dx = -\sum_{j=1}^{n} \int_D \Big(G_j^{\varepsilon}(x) - G_j(x)\Big)\frac{\partial u}{\partial x_j}(x)dx, \end{aligned} \tag{26}$$

since $G_j^{\varepsilon} \equiv G_j$ at ∂D. We suppose that the integrals over ∂D in the second term are absent in the case of $D = \mathbf{R}^n$. Equation (26) implies that

$$|\langle g_{\varepsilon} - g, u\rangle| \leq \sum_{j=1}^{n} \left(G_j^{\varepsilon} - G_j\right)_{L_{\nu}(D)} \|u\|_{\mathcal{W}}$$

for $\nu = \mu(\mu-1)^{-1}$. Hence $g_{\varepsilon} = \sum_{j=1}^{n} \frac{\partial G_j^{\varepsilon}}{\partial x_j}(x) \to g$ in $\mathcal{W}^*$.

It can be easily seen that if D is bounded, then $\mathcal{W}^* \subset H^{-1}$. Furthermore, in a case of $D = \mathbf{R}^n$, we have that there exists a bounded region $D_1 \subset \mathbf{R}^n$ such that $\langle \xi, g\rangle = 0$ $(\forall \xi \in \mathcal{W}:$ $\operatorname{supp}\xi \subset \mathbf{R}^n \backslash D_1$. Hence $g \in H^{-1}$ and $\|g\|_{H^{-1}} \leq c\|g\|_{\mathcal{W}^*}$, where $c = c(n, D_1, \mu)$ is a constant. This completes the proof of Lemma 5.1 in this particular case.

Suppose that $\mathcal{N} = 1$, and the vector $\mathbf{n}_1$ has an arbitrary direction. Introduce new coordinates $\widetilde{x}_j = \sum_{k=1}^{n} \theta_{jk}x_k$, where a matrix $\theta \in \mathbf{R}^{n\times n}$ is such that $\theta^{\top} = \theta^{-1}$, and $|\widetilde{\mathbf{n}}_1^{(j)}| = n^{-1/2}$, $\mathbf{n}_1^{(j)}\widetilde{\mathbf{n}}_1^{(j)} \geq 0$, $j = 1,..,n$, for the corresponding normal in the new coordinates $\widetilde{\mathbf{n}}_1 = (\widetilde{\mathbf{n}}_1^{(1)}, ..., \widetilde{\mathbf{n}}_1^{(n)})$. Let $\widetilde{G}_j(\widetilde{x})$ be the corresponding functions defined as in the formulation of Lemma 5.1, but in the new coordinates. Furthermore, let $\widetilde{g} = \sum_{j=1}^{n} \frac{\partial \widetilde{G}_j}{\partial \widetilde{x}_j}$ and $\widetilde{g}_{\varepsilon}(\widetilde{x}) = \mathbb{I}\{x = \theta^{-1}\widetilde{x} \in \Gamma(\varepsilon)\}$. As above, we have that $\widetilde{g}_{\varepsilon}(\widetilde{x}) \to \widetilde{g}$ in $\mathcal{W}^*$. Let $\Delta G_j(x)$ be the jumps of functions G_j at x on the lines $P_j(x)$, in ascending order of x_j. We have that $\Delta G_j(x) = \gamma_j(x)$. Denote by $\Delta\widetilde{G}_j(\widetilde{x})$ the analogous jumps in the new coordinates, for $\widetilde{\Gamma} = \theta\Gamma$. It is a simple geometric statement, that $\Delta\widetilde{G}_j(\widetilde{x}) = \sum_{k=1}^{n} \theta_{jk}\Delta G_k(\theta^{-1}\widetilde{x})$ if $N(x,j) < +\infty$, $x \in \Gamma\backslash S_0$. Then $\langle\widetilde{\xi}, \widetilde{g}\rangle = \langle \xi, g\rangle$, if $\xi \in \mathcal{W}$, $\xi(x) = \widetilde{\xi}(\theta x)$. This completes the proof of Lemma 5.1 in this case.

Let $\mathcal{N} > 1$, $g_{\varepsilon}^{(i)}(x) = \operatorname{Ind}\{x \in \Gamma_i(\varepsilon)\}$. Then $g_{\varepsilon} = \sum_{i=1}^{\mathcal{N}} g_{\varepsilon}^{(i)}$, $g_{\varepsilon}^{(i)}(x) \to g^{(i)}$ in $\mathcal{W}^*$, $g^{(i)} = \sum_{j=1}^{n} \frac{\partial G_j^{(i)}}{\partial x_j} \in \mathcal{W}^*$, where $G_j^{(i)}$ are such as in the formulation of Lemma 5.1 for corresponding $\Gamma = \Gamma_i$. We have that $G_j(x) = \sum_{i=1}^{\mathcal{N}} G_j^{(i)}(x)$. This completes the proof of Lemma 5.1. □

6 Existence of the Occupation Time Density and an Analog of Meyer-Tanaka Formula

Set

$$g_\varepsilon(x,t) \triangleq \frac{1}{\varepsilon}\mathbb{I}\{x \in \Gamma(\varepsilon,t)\}, \quad l_\varepsilon(t) \triangleq \int_0^t g_\varepsilon(y(s),s).$$

It is natural to interpret the limit of $l_\varepsilon(T)$ as the occupation time of $y(t)$ on $\Gamma(t)$.

Condition 6.1 *(i) The hypersurface $\Gamma(t)$ is such that assumptions of Lemma 5.1 holds for $\Gamma = \Gamma(t)$ for a.e. $t \in [0,T]$ and $g = g(t) \in X^{-1}$, where $g(t) \triangleq \lim_{\varepsilon\to 0} g_\varepsilon(t)$ (by Lemma 5.1, the limit exists in H^{-1} for a.e. $t \in [0,T]$).*

(ii) The initial vector $a = y(0)$ has probability density function $\rho \in L_\infty(\mathbf{R}^n)$.

(iii) The function $\beta(x,t)$ in (1) is continuous.

Note that assumptions of Lemma 5.1 hold for disks, spheres and many other piecewise C^1-smooth $(n-1)$–dimensional surfaces. Moreover, it can be easily found examples when the surface $\Gamma(t)$ changes in time, approaching a fractal, but $g \in X^{-1}$.

Example 6.1 *Let $n = 2$, $T = 2$, $\Gamma(t) = \{\ (x_1,x_2) : x_2 = \sin(x_1(1-t)^{-1/3}), x_1 \in [-1,1]\}$. Then $N((2,x_2),2,\Gamma(t)) \equiv 1$ and*

$$\begin{aligned}\|g\|^2_{X^{-1}} = \textstyle\int_0^T \|g(t)\|_{H^{-1}} dt \quad &\le \text{const} \textstyle\int_0^T \left[1 + \sup_{x_2} N((2,x_2),1,\Gamma(t))^2\right] dt \\ &\le \text{const}\left(2 + \textstyle\int_0^2 (1-t)^{-2/3} dt\right) < +\infty.\end{aligned}$$

Hence $g = g(t) \in X^{-1}$.

The following example present a fractal Γ which constant in time.

Example 6.2 *Let $n = 2$, $\Gamma(t) \equiv \Gamma = \{\ (x_1,x_2) : x_2 = x_1 \sin(x_1^{-1/3}), x_1 \in [-1,1]\}$. Then $N((2,x_2),2,\Gamma(t)) \equiv 1$ and*

$$\|g\|^2_{H^{-1}} \;=\left(1 + \textstyle\int_{-1}^1 dx_2 N((2,x_2),1,\Gamma(t))^2\right) \le \text{const}\left(1 + \textstyle\int_{-1}^1 x_1^{-2/3} dx_1\right) < +\infty.$$

Hence $g \in H^{-1}$.

Denote by β_j the columns of the matrix β, $j = 1,...,n$.

Let $\mathcal{F}_t$ be the filtration of complete σ–algebras of events, generated by $\{a, w(s), \nu(B,s),\ s \le t,\ B \in \bar{\mathcal{B}}_n\}$.

Introduce the set $\widetilde{\mathcal{Y}}$ of all bounded functions $\xi(t) = \xi(t,\omega) : [0,T] \times \Omega \to \mathbf{R}^n$ which are progressively measurable with respect to $\mathcal{F}_t$, and introduce the set $\widetilde{\mathcal{Z}}$ of all bounded functions $\psi(u,t) = \psi(u,t,\omega) : \mathbf{R}^n \times [0,T] \times \Omega \to \mathbf{R}^n$ which are progressively measurable with respect to $\mathcal{F}_t$ for all u.

Introduce the Hilbert space $\mathcal{Y}_2$ as the completion of $\widetilde{\mathcal{Y}}$ with respect to the norm $\|\xi\|_{\mathcal{Y}_2} \triangleq \mathbf{E}\int_0^T |\xi(t)|^2 dt$, and introduce the Hilbert space $\mathcal{Z}_2$ as the completion of $\widetilde{\mathcal{Z}}$ with respect to the norm $\|\psi\|_{\mathcal{Z}_2} \triangleq \mathbf{E}\int_0^T dt \int_{\mathbf{R}^m} |\psi(u,t)|^2 \Pi(du)$.

We present now an analog of the Meyer-Tanaka formula (see e.g. Meyer (1976) or Protter (1990), p.169.

Theorem 6.1 *Assume that Conditions 3.1, 4.1, 4.2 and 6.1 are satisfied. Let* $V \triangleq L(0)g$ *(by the definition, this* V *belongs* Y^1*). Let* V *and* $\partial V/\partial x$ *be Borel measurable representatives* V *and* $\partial V/\partial x$ *of corresponding equivalence classes in* $L_2(Q)$*. Then*

$$\frac{\partial V}{\partial x}(y(t),t)\beta_j(y(t),t) \in \mathcal{Y}_2, \quad V(y(t)+\theta(y(t),u,t),t) - V(y(t),t) \in Z_2,$$

and $\mathbf{E}|l_\varepsilon(T) - \widehat{\mathbf{t}}(T)|^2 \to 0$ *as* $\varepsilon \to 0$*, where*

$$\begin{aligned}\widehat{\mathbf{t}}(T) &\triangleq V(a,0) + \sum_{j=1}^{n}\int_0^T \frac{\partial V}{\partial x}(y(t),t)\beta_j(y(t),t)\,dw_j(t)\\ &+ \int_0^T dt\int_{\mathbf{R}^m}(V(y(t)+\theta(y(t),u,t),t) - V(y(t),t))\widetilde{\nu}(du,dt).\end{aligned} \tag{27}$$

Corollary 6.1 *Let* $p \triangleq \mathbf{L}(0)^*\rho$*. In assumptions and notations of Theorem 6.1,*

$$\begin{aligned}\mathbf{E}\widehat{\mathbf{t}}(T)^2 = \int_{\mathbf{R}^n}|V(x,0)|^2\rho(x)\,dx + \int_Q\Bigg(\sum_{j=1}^{n}\left|\frac{\partial V}{\partial x}(x,t)\beta_j(x,t)\right|^2\\ + \int_{\mathbf{R}^m}|V(x+\theta(x,u,t),t) - V(x,t)|^2\Pi(du)\Bigg)p(x,t)dxdt.\end{aligned}$$

Note that if Condition 6.1(ii) is satisfied, then $\|\rho\|_{H^0}^2 = \int_{\mathbf{R}^n}\rho(x)^2dx \le \|\rho\|_{L_\infty(\mathbf{R}^n)}$, and $\rho \in H^0$. By the definition, p in Corollary 6.1 is the solution of boundary value problem (24) with $\varphi = 0$, $g = 0$. Moreover, by Theorem 4.1, it follows that $p(x,t)$ is the probability density function of the process $y(t)$, and, by Corollary 4.2, $p \in L_\infty(Q)$.

Proof of Theorem 6.1. Set

$$\xi_j(t) \triangleq \frac{\partial V}{\partial x}(y(t),t)\beta_j(y(t),t), \quad \psi(u,t) \triangleq V(y(t)+\theta(y(t),u,t),t) - V(y(t),t).$$

Let $h_\varepsilon(x,t) \in X^0 \cap C([0,T];C^2(\mathbf{R}^n))$ be such that $\|h_\varepsilon - g_\varepsilon\|_{X^0} \le \varepsilon$. Set

$$V_\varepsilon \triangleq L(0)h_\varepsilon, \quad \lambda_\varepsilon(t) \triangleq \int_0^t h_\varepsilon(y(s),s)\,ds,$$

$$\xi_{j,\varepsilon}(t) \triangleq \frac{\partial V_\varepsilon}{\partial x}(y(t),t)\beta_j(y(t),t),$$

$$\psi_\varepsilon(u,t) \triangleq V_\varepsilon(y(t)+\theta(y(t),u,t),t) - V_\varepsilon(y(t),t).$$

By the definition, we have that $h_\varepsilon = -\partial V_\varepsilon/\partial t - \mathcal{A}V_\varepsilon$ and $V_\varepsilon(x,T) = 0$. By the Generalized Itô formula (see e.g. Gikhman and Skorohod (1972), p.272), it follows that

$$\begin{aligned}-V_\varepsilon(a,0) &= V_\varepsilon(y(T),T) - V_\varepsilon(a,0)\\ &= -\int_0^T h_\varepsilon(y(t),t)\,dt + \sum_{j=1}^{n}\int_0^T \xi_j(y(t),t)\,dw_j(t)\\ &+ \int_0^T dt\int_{\mathbf{R}^m}(V_\varepsilon(y(t)+\theta(y(t),u,t),t) - V_\varepsilon(y(t),t))\widetilde{\nu}(du,dt).\end{aligned}$$

Hence

$$\lambda_\varepsilon(T) = V_\varepsilon(a,0) + \sum_{j=1}^{n}\int_0^T \frac{\partial V_\varepsilon}{\partial x}(y(t),t)\beta_j(y(t),t)\,dw_j(t)$$
$$+ \int_0^T dt\int_{\mathbf{R}^m}(V_\varepsilon(y(t)+\theta(y(t),u,t),t) - V_\varepsilon(y(t),t))\widetilde{\nu}(du,dt).$$

By Lemmas 2-3 and Theorem 4 from Gikhman and Skorohod (1972) pp. 293-296, it follows that the functions V_ε and $\partial V_\varepsilon(x,t)/\partial x$ are bounded and continuous, then $\xi_{\varepsilon,j}(t) \in \mathcal{Y}^0$ and $\psi_\varepsilon(u,t) \in \mathcal{Z}^0$.

By Theorem 4.1, $p = p(x,t) \stackrel{\Delta}{=} \widehat{\mathbf{L}}(0)^*\rho$ is the probability density function of the process $y(t)$. Let $W_\varepsilon \stackrel{\Delta}{=} V_\varepsilon - V$. By Corollary 3.1, $\|W_\varepsilon\|_{Y^1} \to 0$ as $\varepsilon \to 0$. Then

$$\mathbf{E}|V_\varepsilon(a,0) - V(a,0)|^2 = \int_{\mathbf{R}^n}\|W_\varepsilon(x,0)\|^2\rho(x)dx \le \|\rho\|_{L_\infty(\mathbf{R}^n)}\|W_\varepsilon\|_{X^0} \to 0,$$

$$\mathbf{E}\int_0^T|\xi_{j,\varepsilon}(t) - \xi_j(t)|^2dt = \sum_{j=1}^{n}\mathbf{E}\int_0^T\left|\frac{\partial W_\varepsilon}{\partial x}(y(t),t)\beta_j(y(t),t)\right|^2dt$$
$$= \int_Q\left|\frac{\partial W_\varepsilon}{\partial x}(x,t)\right|^2 p(x,t)\,dx\,dt \le \text{const}\,\|p\|_{L_\infty(Q)}\|W_\varepsilon\|_{Y^1} \to 0,$$

and

$$\mathbf{E}\int_0^T dt\int_{\mathbf{R}^n}|\psi_\varepsilon(u,t) - \psi(u,t)|^2\Pi(du) = \mathbf{E}\int_0^T|(\mathcal{J}(t)W_\varepsilon)(y(t),t)|^2dt$$
$$= \int_Q|(\mathcal{J}(t)W_\varepsilon)(x,t)|^2 p(x,t)dxdt \le \text{const}\,\|p\|_{L_\infty(Q)}\|\mathcal{J}W_\varepsilon\|_{X^0} \to 0.$$

Then

$$\mathbf{E}|\widehat{\mathbf{t}}(T) - \lambda_\varepsilon(T)|^2 = \mathbf{E}|V_\varepsilon(a,0) - V(a,0)|^2 + \sum_{j=1}^{n}\mathbf{E}\int_0^T|\xi_{j,\varepsilon}(t) - \xi_j(t)|^2dt$$
$$+\mathbf{E}\int_0^T dt\int_{\mathbf{R}^n}|\psi_\varepsilon(u,t) - \psi(u,t)|^2\Pi(du) \to 0. \qquad (28)$$

Furthermore,

$$\mathbf{E}|l_\varepsilon(T) - \lambda_\varepsilon(T)|^2 = \int_0^T\int_{\mathbf{R}^n}|g_\varepsilon(x,t) - h_\varepsilon(x,t)|^2p(x,t)dt \le \|p\|_{L_\infty(Q)}\|g_\varepsilon - h_\varepsilon\|_{X^0} \to 0.$$

This completes the proof of Theorem 6.1. □

Proof of Corollary 6.1 is can be easily obtained similarly (28).

7 Equations for the Characteristic Function of the Occupation Time

Theorem 7.1 *Assume that Conditions 3.1, 4.1, 4.2 and 6.1 are satisfied, and that $g \in X$. Let $\nu \in \mathbf{R}$ be given, and let $z \stackrel{\Delta}{=} i\nu$, where $i = \sqrt{-1}$. Let $V \stackrel{\Delta}{=} zL(zg)g$. Then $V \in Y^1$, and*

$$1 + (V(\cdot,0),\rho)_{H^0} = \mathbf{E}\exp\{z\widehat{\mathbf{t}}(T)\}. \qquad (29)$$

Proof. Let h_ε be such that $h_\varepsilon(x,t) \in X^0 \cap C([0,T];C^2(\mathbf{R}^n))$ and $\|h_\varepsilon - g_\varepsilon\|_X \le \varepsilon$. Set $\lambda_\varepsilon(t) \triangleq \int_0^t h_\varepsilon(y(s),s)ds$. Let

$$V_\varepsilon(x,s) \triangleq \mathbf{E}\left\{ z\int_s^T h_\varepsilon\Big(y(t),t\Big)\exp\left(z\int_s^t h_\varepsilon\Big(y(r),r\Big)dr\right)dt \,\Big|\, y(s)=x\right\}.$$

It is easy to see that

$$V_\varepsilon(x,s) = \mathbf{E}\left\{\exp\left(z\int_s^T h_\varepsilon\Big(y(t),t\Big)dt\right)\Big|\, y(s)=x\right\} - 1.$$

By Theorem 1 of Gikhman and Skorohod (1972), p.301, applied after a small modification for a non-homogenuous integro-differential equation, it follows that $V_\varepsilon = zL(zh_\varepsilon)h_\varepsilon$, i.e. V_ε is the solution of the problem

$$\begin{cases} \frac{\partial V_\varepsilon}{\partial t} + \mathcal{A}V_\varepsilon + zh_\varepsilon V_\varepsilon = -zh_\varepsilon, \\ V_\varepsilon(x,T) = 0. \end{cases} \tag{30}$$

By Lemma 5.1 it follows that $\|g-g_\varepsilon\|_X \to 0$ as $\varepsilon \to 0+$. Hence $\|g-h_\varepsilon\|_X \to 0$ as $\varepsilon \to 0+$. By Corollary 3.1, it follows that $\|V - V_\varepsilon\|_{Y^1} \to 0$. Hence $(V_\varepsilon(\cdot,0),\rho)_{H^0} \to (V(\cdot,0),\rho)_{H^0}$. It was shown in the proof of Theorem 6.1 that $\mathbf{E}|\lambda_\varepsilon(T) - \widehat{\mathbf{t}}(T)|^2 \to 0$. Then $\lambda_\varepsilon(T)$ converge to $\widehat{\mathbf{t}}(T)$ in distribution, and $\mathbf{E}e^{z\lambda_\varepsilon(T)} \to \mathbf{E}e^{z\widehat{\mathbf{t}}(T)}$ for each $z = i\nu$, $\nu \in \mathbf{R}$. Then the proof follows. □

8 Variations of Functionals of Occupation Time

Assume that all the assumptions and notations are valid. In particular, we assume that a $(n-1)$–surface Γ is such that assumptions of Lemma 5.1 hold, $g \in X$ is defined in Lemma 5.1, $y(t)$ is the solution of (1), the values $\widehat{\mathbf{t}}(t)$ are defined in Theorem 6.1, the initial vector a has the probability density function $\rho \in H^0$.

For $\delta \in [0,1]$, we consider a set of functions $f = f_\delta : Q \to \mathbf{R}^n$, $\beta = \beta_\delta : Q \to \mathbf{R}^{n\times n}$, $\Pi = \Pi_\delta : \bar{\mathcal{B}}_m \to \mathbf{R}$, such that all the assumptions of Section 2 hold for all δ and

$$f_\delta(x,t) = f_0(x,t) + \delta\widehat{f}(x,t), \quad \beta_\delta(x,t) = \beta_0(x,t) + \delta\widehat{\beta}(x,t), \quad \Pi_\delta(du) = \Pi_0(du) + \delta\widehat{\Pi}(du),$$

where $\widehat{f}(x,t)$, $\widehat{\beta}(x,t)$ are bounded and Borel measurable functions, and $\widehat{\Pi}(du)$ is a measure on $\mathbf{R}^m$ such that $\widehat{\Pi}(\mathbf{R}^m) < +\infty$. We assume that the derivative $\partial\widehat{\beta}(x,t)/\partial x$ is bounded.

Let $c_1 \in \mathbf{R}, c_2 \in \mathbf{R}, z \in \mathbf{C}$ be constants. Let

$$\Phi_\delta = c_1\mathbf{E}\widehat{\mathbf{t}}(T) + c_2\mathbf{E}e^{z\widehat{\mathbf{t}}(T)}, \quad \delta \in [0,1]$$

be values defined for $(f,\beta,\Pi) = (f_\delta,\beta_\delta,\Pi_\delta)$.

For the sake of simplicity, we suppose that either $(c_1,c_2) = (0,1)$ or $(c_1,c_2) = (1,0)$.

Let

$$V_\delta = c_1 L(0)g + c_2(zL(zg)g + 1), \qquad p_\delta = c_1\widehat{\mathbf{L}}(0)^*\rho + c_2\widehat{\mathbf{L}}(zg)^*\rho$$

be elements of Y^1 defined for $(f,\beta,\Pi) = (f_\delta,\beta_\delta,\Pi_\delta)$, and for the operators $L(0), L(zg), \widehat{\mathbf{L}}(0)^*, \widehat{\mathbf{L}}(zg)^*$ introduced in Sections 3, 4.

We introduce the matrix function $B \triangleq \frac{1}{2}(\beta_0\widehat{\beta}^\top + \widehat{\beta}\beta_0^\top)$.

Let $\mathcal{A}_\delta$ be operator (6) defined for $(f,\beta,\Pi) = (f_\delta,\beta_\delta,\Pi_\delta)$.

Theorem 8.1 *There exists the limit*

$$\begin{aligned}\lim_{\delta\to 0+}\frac{\Phi_\delta-\Phi_0}{\delta} &= \int_Q p_0(x,t)\frac{\partial V_0}{\partial x}(x,t)\widehat{f}(x,t)\,dx\,dt\\ &- \sum_{i,j=1}^{n}\int_Q \frac{\partial}{\partial x_i}(p_0(x,t)B_{ij}(x,t))\frac{\partial V_0}{\partial x_j}(x,t)\,dx\,dt\\ &+ \int_Q p_0(x,t)\int_{\mathbf{R}^m} V_0(x+\theta(x,u,t),t)\widehat{\Pi}(du)\,dx\,dt. \end{aligned}\tag{31}$$

Lemma 8.1 *There exists $\delta_*\in(0,1)$ such that the following estimate holds: $\|V_\delta - V_0\|_{Y^1}\le c_0\delta$ $(\forall\delta\in[0,\delta_*))$, where a constant c_0 does not depend on δ.*

Proof. Let $b_\delta=\frac{1}{2}\beta_\delta\beta_\delta^\top$, $W=V_\delta-V_0$. We have that

$$\begin{cases}\frac{\partial W}{\partial t}+\mathcal{A}_\delta W+c_2 zgW=-\xi,\\ W(x,t)|_{x\in\partial D}=0,\quad W(x,T)=0,\end{cases}\tag{32}$$

where

$$\begin{aligned}\xi(x,t)\stackrel{\Delta}{=}(\mathcal{A}_\delta-\mathcal{A}_0)V_0(x,t) &= \frac{\partial V_0}{\partial x}(x,t)(f_\delta(x,t)-f_0(x,t))\\ &- \sum_{i,j=1}^{n}(b_s-b_0)_{ij}(x,t))\frac{\partial^2 V_0}{\partial x_i\partial x_j}(x,t)\\ &+ \int_{\mathbf{R}^m} V_0(x+\theta(x,u,t),t)[\Pi_\delta(du)-\Pi_0(du)].\end{aligned}\tag{33}$$

Clearly, $f_\delta-f_0=\delta\widehat{f}$, $b_\delta-b_0=\delta B+\delta^2\widehat{\beta}\widehat{\beta}^\top$, and $\Pi_\delta-\Pi_0=\delta\widehat{\Pi}$. Then

$$\|\xi\|_{X^{-1}}\le\delta\widehat{c}_1\|V_0\|_{X^1}\quad\forall\delta\in[0,1).$$

By Lemma 3.1, we have that

$$\|W\|_{Y^1}\le\widehat{c}_2\|\xi\|_{X^{-1}}\le\delta\widehat{c}_3\|V\|_{X^1}\le c_0\quad\forall\delta\in[0,\delta_*)$$

for small enough $\delta_*>0$. The constants $c_0,\widehat{c}_1,\widehat{c}_2,\widehat{c}_3$ do not depend on δ. This completes the proof of Lemma 8.1.□

Lemma 8.2 *The following equality holds:*

$$\begin{aligned}\Phi_\delta-\Phi_0 &= \int_Q p_0(x,t)\Big(\mathcal{A}_\delta(x,t)-\mathcal{A}_0(x,t)\Big)V_0(x,t)\,dx\,dt\\ &+ \int_Q p_0(x,t)\Big(\mathcal{A}_\delta(x,t)-\mathcal{A}_0(x,t)\Big)\Big(V_\delta(x,t)-V_0(x,t)\Big)\,dx\,dt.\end{aligned}\tag{34}$$

Proof. Let the sets $\Gamma(\varepsilon)$ be defined by (3), $\varepsilon\to 0+$, $g_\varepsilon(x)=\varepsilon^{-1}\mathbb{I}\{x\in\Gamma(\varepsilon)\}$. Let $V_\delta^\varepsilon=c_1L(0)g_\varepsilon+c_2(zL(zg_\varepsilon)g_\varepsilon+1)$, $p_\delta^\varepsilon=c_1\widehat{\mathbf{L}}(0)^*\rho+c_2\widehat{\mathbf{L}}(zg_\varepsilon)^*\rho$ be elements of Y^1 defined for the corresponding coefficients $f=f_\delta$, $\beta=\beta_\delta$ and for the operators $L(0),L(zg),\widehat{\mathbf{L}}(0)^*,\widehat{\mathbf{L}}(zg)^*$

introduced in Sections 3, 4. We have that $V^\varepsilon_\delta \in Y^2$. Let $\widehat{\mathcal{A}}_\delta = \mathcal{A}_\delta + c_2 z g_\varepsilon I$, $\varphi = c_1 g_\varepsilon + c_2 z g_\varepsilon$. By Theorem 6.1, it follows that

$$\begin{aligned}
\Phi_\delta - \Phi_0 &= (V^\varepsilon_\delta(\cdot,0),\rho)_{H^0} - (V^\varepsilon_0(\cdot,0),\rho)_{H^0} = (V^\varepsilon_0(x,T),p^\varepsilon_0(x,T))_{H^0} \\
&- (V^\varepsilon_0(x,0),p^\varepsilon_0(x,0))_{H^0} - (V^\varepsilon_\delta(x,T),p^\varepsilon_0(x,T))_{H^0} + (V^\varepsilon_\delta(x,0),p^\varepsilon_0(x,0))_{H^0} \\
&= \int_0^T \left\{ \left(V^\varepsilon_0, \frac{\partial p^\varepsilon_0}{\partial t}\right)_{H^0} + \left(\frac{\partial V^\varepsilon_0}{\partial t}, p^\varepsilon_0\right)_{H^0} - \left(V^\varepsilon_\delta, \frac{\partial p^\varepsilon_0}{\partial t}\right)_{H^0} + \left(\frac{\partial V^\varepsilon_\delta}{\partial t}, p^\varepsilon_0\right)_{H^0} \right\} dt \\
&= \int_0^T \Big\{ (\widehat{\mathcal{A}}_0 V^\varepsilon_0, p^\varepsilon_0)_{H^0} + (-\widehat{\mathcal{A}}_0 V^\varepsilon_0 - \varphi, p^\varepsilon_0)_{H^0} - (\widehat{\mathcal{A}}_0 V^\varepsilon_\delta, p^\varepsilon_0)_{H^0} \\
&+ (\widehat{\mathcal{A}}_\delta V^\varepsilon_\delta + \varphi, p^\varepsilon_0)_{H^0} \Big\} dt = \int_0^T \left\{ (\widehat{\mathcal{A}}_\delta V^\varepsilon_\delta, p^\varepsilon_0)_{H^0} - (\widehat{\mathcal{A}}_0 V^\varepsilon_\delta, p^\varepsilon_0)_{H^0} \right\} dt.
\end{aligned}$$

It can be easily seen that

$$\begin{aligned}
(\widehat{\mathcal{A}}_\delta V^\varepsilon_\delta, p^\varepsilon_0)_{H^0} - (\widehat{\mathcal{A}}_0 V^\varepsilon_\delta, p^\varepsilon_0)_{H^0} &= (\widehat{\mathcal{A}}_\delta V^\varepsilon, p^\varepsilon_0)_{H^0} - (\widehat{\mathcal{A}}_0 V^\varepsilon_0, p^\varepsilon_0)_{H^0} \\
&+ ((\widehat{\mathcal{A}}_\delta - \widehat{\mathcal{A}}_0)(V^\varepsilon_\delta - V^\varepsilon_0), p^\varepsilon_0)_{H^0}.
\end{aligned}$$

It was proved in Section 3 that $V^\varepsilon_\delta \to V_\delta$ in Y^1 as $\varepsilon \to 0$; similarly, $p^\varepsilon_\delta \to p_\delta$ in Y^1. This completes the proof of Lemma 8.2. □

Proof of Theorem 8.1. By Lemma 8.2, it follows that

$$\begin{aligned}
\frac{\Phi_\delta - \Phi_0}{\delta} &= \int_Q p_0(x,t) \frac{\partial V_0}{\partial x}(x,t) \widehat{f}(x,t)\, dx\, dt \\
&- \sum_{i,j=1}^n \int_Q \frac{\partial}{\partial x_i}(p_0(x,t) B_{ij}(x,t)) \frac{\partial V_0}{\partial x_j}(x,t)\, dx\, dt \\
&+ \int_Q p_0(x,t) \int_{\mathbf{R}^m} V_0(x+\theta(x,u,t),t) \widehat{\Pi}(du)\, dx\, dt + \eta(\delta),
\end{aligned}$$

where

$$\eta(\delta) \stackrel{\Delta}{=} -\delta \sum_{i,j=1}^n \int_Q \frac{\partial}{\partial x_i}\left(p_0(x,t)(\widehat{\beta}\widehat{\beta}^\top)_{ij}(x,t)\right) \frac{\partial V_0}{\partial x_j}(x,t)\, dx\, dt + \eta_1(\delta),$$

and where $\eta_1(s)$ is such that

$$|\eta_1(\delta)| \le c_* \|p_0\|_{Y^1} \|V_\delta - V_0\|_{Y^1} \sup_{x,t}(|\beta(x,t)|^2 + |\widehat{\beta}(x,t)|^2 + |\widehat{f}(x,t)| + 1)^2,$$

where a constant c_* does not depend on δ. By Lemma 8.1, $|\eta_1(\delta)| = o(\delta)$. This completes the proof of Theorem 8.1.

References

[1] Bass, R.F. 1983. Occupation time of d–dimensional semimartingales. *Seminar on Stochastic Processes. Progr. Prob. Stat.* **5** (1983), 51–96.

[2] Benes, V.E, and Shepp, L.A. Some solvable stochastic control problems. *Stochastics* **4** (1980), 134-160.

[3] Borodin, A.N. Brownian local time. 1989. *Russian. Math. Surveys* **44** (2), 1–51.

[4] Carrany, G.M. and Menaldy, J.L. 1992. *Green functions for second order integro-diffferential problems.* Pitman Research Notes in Math. Series.

[5] Davis, M.H.A., and A.R. Norman A.R., Portfolio selection with transaction costs. *Mathematics of Operation Research* **15** (1990), 676-713.

[6] Dokuchaev, N.G. 1999. Local sojourn time of diffusion and degenerating processes on a mobile surface. *Th. Probab. Appl.* **43**, N2, 171-188.

[7] Geman, D., Horovitz, G. 1980. Occupation density. *The Annals of Probability.* **8**, 1–67.

[8] Gikhman, I.I., Skorohod, A.V. 1972. *Stochastic differential equations.* Berlin, Springer.

[9] Harrison, J.M., and Taylor, A.J. Optimal control of a Brownian Storage Systems. *Stochastic Processes Appl.* **6** (1978), 179-194.

[10] Karatzas, I., and Shreve, S.E. Connection between optimal stopping and singular stochastic control. I, II. *SIAM J. Control Optimization* **22** (1984), 856-877 and **23** (1985), 433-451.

[11] Ladyzenskaya, O.A., Solonnikov, V.A., Ural'ceva, N.N. 1968. *Linear and Quasi–Linear Equations of Parabolic Type.* Providence, R.I.: American Mathematical Society.

[12] McGill, P. 1982. Markov properties of diffusion local time: a martingale approach. *Adv. in. Appl. Probab.*. **14**, 789-810.

[13] Meyer, P.A. 1976. *Un cours sur les intégrales stochastiques.* Séminaire Proba. X. Lecture Notes in Mathematics, vol. 511, 246-400. Springer. Berlin Heidelberg New York.

[14] Nolan, J. 1989. Local nondeterminism and local time for stable processes. *Probab. Th. Relat. Fields.*. **82**, 387-310.

[15] Protter, P. 1990. *Stochastic Integration and Differential Equations,* Springer Verlag.

[16] Rosen, J.S., Yor, M. 1991. Tanaka formulae and renormalization for triple intersections of Brownian motion in the plane. *The Annals of Probability.* **19**, 142-159.

[17] Salminen, P. 1993. On the distribution of supremum of diffusion local time. *Statist. Probab. Letters.*. **18**, 219-225.

INDEX

A

B

C

D

E

F

G

H

I

Q

R

S

T